Pedro Pablo Magaña Herrera

Evolution of the Tendon Wall Construction System

Pedro Pablo Magaña Herrera

Evolution of the Tendon Wall Construction System

from Cardellach to Thomas

ScienciaScripts

Imprint

Cover image: www.ingimage.com

This book is a translation from the original published under ISBN 978-620-0-01106-0.

Publisher:
Sciencia Scripts
is a trademark of
Dodo Books Indian Ocean Ltd. and OmniScriptum S.R.L publishing group

120 High Road, East Finchley, London, N2 9ED, United Kingdom
Str. Armeneasca 28/1, office 1, Chisinau MD-2012, Republic of Moldova, Europe
Managing Directors: Ieva Konstantinova, Victoria Ursu
info@omniscriptum.com

Printed at: see last page
ISBN: 978-620-8-52477-7

EVOLUTION OF THE TENDON WALL CONSTRUCTION SYSTEM, FROM CARDELLACH TO THOMAS

PEDRO PABLO MAGAÑA HERRERA

2024

TABLE OF CONTENTS

SUMMARY..3

INTRODUCTION..4

CHAPTER I..5

CHAPTER II..26

CHAPTER III...38

CHAPTER IV...50

BIBLIOGRAPHICAL REFERENCES...................................60

SUMMARY

This research integrates the existing relationship between construction technology and the environment, to be used in the construction of houses using the system of non-structural tendon walls, where this new non-traditional construction technology is presented, as it integrates materials of regional origin and low ecological impact, with the aim of achieving constructive sustainability at environmental, economic and social levels. This construction system is primarily intended to meet the requirements of safety and structural strength, thus becoming a basic tool that should ensure a decent quality of life for all its occupants, responding to the unmet primary needs of families who are vulnerable from the economic point of view, which is to have a roof that provides shelter and can develop their socio-cultural activities. Theoretically, Cardellach's philosophical principle incorporates the term sinewy systems into structures as building forms that have their origin in the zoological architecture of vertebrates, where these structural and building forms are at a higher level of mechanical sensibility and natural inspiration. In the 1990s, Thomas integrated design, technology and culture in an unconventional construction alternative called tendon walls. The typology of this construction system consists of the on-site fabrication of flat rectangular modular panels, reinforced on the inside with a barbed wire mesh that serves as integrated tendons, covered on both sides with a mortar mixture, to form a rigid structural frame, composed of one of the following types of materials: sawn timber, by bamboo or guadua angustifolia (Colombia), by metallic angle or by concrete structure; thus constituting a monolithic structural element that will behave as a confined wall, which will give consistency and finish to the wall.

Key words: non-conventional construction techniques, sustainable environment, sinewy wall, unreinforced masonry.

INTRODUCTION

The construction industry is currently going through a great moment, not only using its traditional construction methods, but within its evolutionary cycle it is implementing new ecological systems that are environmentally friendly. The investigated model consists of a new construction system, composed of light tendinous walls for non-structural use, and is framed in the field of Materials Engineering and Construction Techniques, and is approached from the environmental point of view in area of recycling materials for construction, It is necessary to search for alternatives and mainly for non-traditional raw materials, with the aim of proposing technological solutions that optimise the use of these available resources, which results in the wellbeing of the community, from an economic point of view, and the improvement of the environment.

Chapter I provides a historical summary of the constructive antecedents made by human beings, from the first beginnings of the palafitic constructions, following later with the masonry in the use of stone, earth and adobe as constructive material for their shelters against the phenomena of nature.

The Mesopotamian civilisations promoted and developed their constructions based on masonry, mainly in adobe and later in brick, with the Roman civilisation came the innovation of the masonry construction system, using Roman concrete, composed mainly of lime, water and volcanic ash or pozzolana, giving resistance and durability over time, leaving constructions that endure to the present day. In the 19th century, cement was invented in England, composed of a mass of clayey limestone and coal, which were cooked at high temperatures, giving as a product a mixture called clinker, native to Portland, when mixed with sand, crushed stone and water, it was given the name of concrete. Later in the middle of the same century, concrete was invented. reinforced concrete, consisting of steel bars and concrete mix, this construction technique is still prevalent today. In addition, a retrospective is made organic, inorganic and composite building materials.

Chapter II gives an overview of sustainable building materials, starting with their life cycle, the prospects of the circular economy in the field of construction, and the environmental problems of building materials.

In Chapter III, a brief historical review is made of the principle of tendon structures, starting with the creator of this principle, Eng. Félix Cardellach, analysing this new subject as a non-conventional construction system and its components.

In Chapter IV, the contextualisation the different analyses and points of view made by a number of researchers interested in this new constructive subject is presented for information purposes. Finally, the discussion and conclusions of the tendon wall construction system are presented.

Following these environmentalist paths, this research is carried out with the purpose of contributing and presenting to the knowledge, a new construction system, taking into account from this perspective, the application of new clean technologies in the field of construction of housing of all kinds and mainly of social interest (VIS).

CHAPTER I.

CONSTRUCTIVE BACKGROUND

1.1.- Historical Overview. According to historical chronology, the zoological family Hominidae, which includes the human genus, is one of the families of the order Primates, one of the seventeen higher groupings generally recognised among the existing placental mammals (Howell, 1982, p. 21-22). Approximately ten million years ago, these hominids in their evolutionary process were herbivores and nomads who travelled great distances daily in search of plant foods. Due to the prevailing climate, they sought refuge and security in the treetops, at the foot of mountains and cliffs; if the climate was very cold, they sought natural cavities called caverns, where they found shelter and refuge from the rain and prevailing winds (Comas, 1977, p. 58). In order to take shelter from the weather, the rain and particularly from wild animals, the cave man began his construction phase, when he built the first stone walls at the entrance of the caves (Perdrizet, 1987, p. 24), this construction method is called masonry, from the Latin manus-positus (to place with the hand).

Following the evolutionary, biological and cultural process of the human being, our prehistoric ancestors elaborated stone tools as weapons, to be used in hunting and fishing animals, adding proteins to their food diet, this historical period is called Palaeolithic. With the climatic changes of the last glaciation about twelve thousand years BC, in the region extending from Palestine, through Syria to Mesopotamia, called the fertile crescent, caused geographical, social and cultural changes that allowed the beginning of new activities such as agriculture by cultivating plants and livestock by domesticating animals for their own benefit. The transformation of some groups from hunters, fishermen and gatherers to farmers, and from nomadic to sedentary, constituted a decisive revolution in human history, (De Roux, 1990, p. 26-27). This new way of life had drastic influence on social relationships, giving rise to to community life, forming small villages made up of lake dwellings or palafittes in some . Comas affirms that the palafittes correspond to very advanced periods of prehistory, beginning the Neolithic, since their construction implies sedentarism (1977, p. 60).

These palafittes were huts built with wooden stakes and lianas from trees, erected at a certain distance from the shore of a river, lake or sea, the construction method was to drive the stakes, poles or posts into the bottom of the water at a certain distance, with lengths of between three and six metres, They were stiffened or hardened by fire to extract the sap, thus extending their lifespan. Later, the intermediate part of the posts were joined with horizontal crosspieces forming a trellis, on which a kind of platform or floor was built at a certain height above the water level to prevent flooding. This kind of wooden trellis was made laterally until it reached the heads of the poles, forming the walls, and the roof was made with the branches of the trees or palms (Comas, 1977, p.183), Figure 1.1.

Figure No. 1.1
Construction of stilt houses

Source:
https://es.wikipedia.org/wiki/Sitios_palaf%C3%ADticos_prehist%C3%B3ricos_de_los_Alpes

In arid geographical regions where wood is scarce, but reeds are plentiful, huts were built with tall and strong elements that were well tied together and used as parales that constituted the main framework, incorporating more reeds to make the framework of the walls that The roofs were then covered entirely with earth or wet mud, and exactly the same construction technique was used for the roofing. The roofs that were unable to withstand winter storms were replaced by double pitched roofs, and thus, by covering the sloping roofs with mud, they were able to allow the rainwater to run off (Vitruvii, 1997, p. 54).In order to protect himself from the weather, the rain and particularly from wild animals, the cave man began his construction phase when he built the first stone walls at the entrance to the caves (Perdrizet, 1987, p. 24), this construction method is called masonry, from the Latin manus-positus (to place by hand). Approximately 8000 years ago a. C., in some zones or areas where agriculture was highly developed, an unprecedented urban revolution took place, marking the beginning of a new milestone or new phase in the history of mankind, generating much more complex societies than those known historically, these were the first civilisations, whose etymology comes from the Latin words civitas (city), civue (citizen), which as a whole translates as urbanised society. The construction also underwent evolution, the walls were built using a set of geometric pieces called adobes, this new construction material is composed of a mass of mud or clay or silt, sand and water, mixed with chopped fragments of dry grass or straw (dry stalks of grass plants) used as reinforcement of the mixture, The adobe pieces were moulded geometrically with wooden slats and then left to bake or dry in the open air, initiating the construction process. These adobe pieces were then glued together with a kind of mortar composed of the same materials used to make the adobes, but with more straw added to the mixture.2.Vitruvii, (1997), states that: "They should be made neither of sand nor of stony or coarse sandy soil, because if they are made of these soils they are heavy and when they are placed on the walls, they decompose due to the effect of the rain and fall apart, and the straws do not bind well due to their roughness", (p. 58). This knowledge made it possible to manipulate the edaphic materials to modify their properties by adding a series of materials of lithological and organic origin, thereby controlling and stabilising different properties inherent to the material itself (Gama et al., 2012).

Figure No. 1.2
Mud brick construction, Niswa (Oman)

Source: https://stock.adobe.com/es/images/edificaciones-antiguas-en-adobe-nizwa-oman/284465081

The great disadvantages of adobe constructions was its low resistance to water attack, to avoid this problem, adobe began to be fired at high temperatures and was transformed into brick, Torroja, (2010) states that being: "the first material created by the dominion of human intelligence over the four elements: earth, air, water and fire, that material so docile and human in which the mud after laborious kneading, skillful moulding and patient drying became stone in the heat of a fire. One of the main constructive characteristics of brick is its hardness and resistance, which is why in its elaboration they dispensed with the straw that was a necessary addition to dry the adobes, initiating the constructive process these pieces of brick were glued with mortar, from the Latin caementum, a kind of mortar composed slaked lime, sand and water, using the same constructive techniques of the adobe, managing to perfect their cemented houses, they built brick walls. or stone, with various kinds of wood they built their roofs and covered them with tiles, (Vitruvii, 1997). With this new material and improving their construction techniques, man developed countless projects in masonry such as houses for dwellings, temples for their religious worship, palaces for their rulers and walls as defensive structures to protect their cities from enemies, therefore it became an essential material for this new civilised world, being one of the oldest materials in the construction world invented by man that still prevails to this day.

Figure No. 1.3

Brick Religious Complex, San Francisco Church, Cali

Source:
https://es.wikipedia.org/wiki/Complejo_religioso_de_San_Francisco_%28Cali%29

With the passing of the centuries, man continued to build great constructions using stone or clay materials such as masonry, giving them stability with the mixture of mortar, these mixtures were originally developed and came from their geographical location. An example of this kind of constructions are their prodigious constructions of religious temples, (Figure 1.3), palaces and their famous funerary constructions that still survive like the Egyptian pyramids that date around 2,570 years B.C., where pastes were used as mortar. C., where pastes obtained from mixtures of gypsum and limestone dissolved in water were used to solidly join stone ashlars (Vidaud, 2013). Greek builders 500 BC used a mortar made of limestone, sand and water, giving good levels of resistance to their monumental buildings. But it was the Romans in the 2nd century B.C. who invented a new mortar called pozzolanic cement, composed of calcined limestone, fine sand of volcanic origin, used in the construction of the Roman Colosseum (Figure 1.4), and the Theatre of Pompeii. Vidaud (2013), adds that: "pozzolana contains silica and alumina, which when combined with lime results in pozzolanic cement; a material that has proven to have great performance, both in terms of strength and durability" (p. 21).

Figure No. 1.4

Construction of the Roman Colosseum with pozzolanic concrete

Source: https://es.wikipedia.org/wiki/Coliseo

In Puzol, a municipality located on the slopes of Vesuvius, there is a kind of powder that contains true wonders in a natural way, mixed with lime and rough stone or pieces of brick, it offers a great solidity to buildings and constructions that are made under the sea, as it consolidates under water (Vitruvii, 1997). With the fall of the Roman Empire, its use declined sharply and most of the building knowledge almost completely disappeared. For a long time, lime containing clays was considered unsuitable for the manufacture of limes and mortars. In the 18th century in England, it was found that some limes made with limestone and clay produced mortars that were more resistant than those made with pure limes and with a supremely important property: they set in the presence of water, which was not the case with the traditional mortars used up to that time (Arredondo, 1969). In 1845 Isaac Johnson improved the process previously patented by Aspdin & Parker by obtaining a product called Clinker and called it Portland Cement, because of the similarity in colour of some limestone rocks from the quarries on that island. The new Portland cement is a synthetic material (artificial rock) that sets and hardens by means of a chemical reaction with water, product of the controlled calcination at high temperatures of clayey materials and limestone, for (Keyser, 1982), "the clayey materials provide silica (SiO_2) and the calcined limestones calcium silicates (CaSiO3), for this reason it is called hydraulic cement", (p. 298). The invention of reinforced concrete is attributed to the builder William Boutland Wilkinson, who in 1854 applied for a patent for a concrete construction system that included iron reinforcement, the technique he used consisted of a formwork with plaster casings and casting concrete on iron bars, forming a ribbed slab, (Valenzuela, 2015, p. 135), becoming the first builder to place iron reinforcement in the ribs of the slab. 135), becoming the first builder to place iron reinforcement inferior to flexural tension in the ribs of the slab, he later built the first reinforced concrete structure, which consisted of a two-storey house made of a mixture of concrete with iron and wire reinforcement in Newcastle, Figure 1.5.

Figure No. 1.5

Reinforced concrete ribbed slab construction

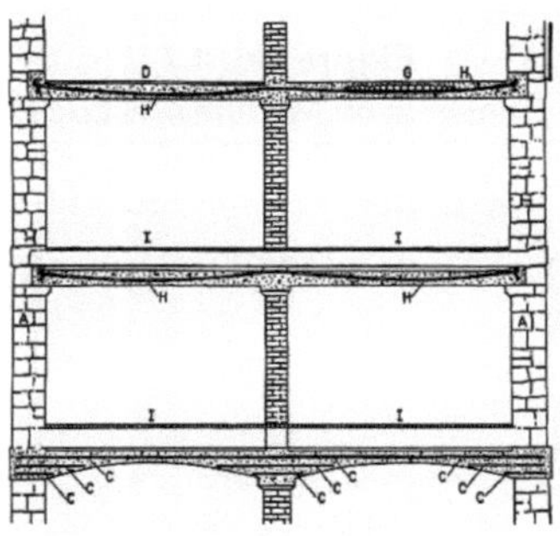

Source: Valenzuela, p.135

Françoise Hennebique in 1892 made an important contribution by using cylindrical section bars, which could be bent in the form of stirrups and used as anchors (Lamuz and Andrade, 2015, p. 29). The structural system used consisted of using a self-supporting skeleton of columns, beams, added with iron rods with stirrups that allowed the support in the right position, and the whole set was cast with a dosed mixture of concrete, Figure 1.6.

Figure No. 1.6

Construction of columns and ribbed slabs in reinforced concrete

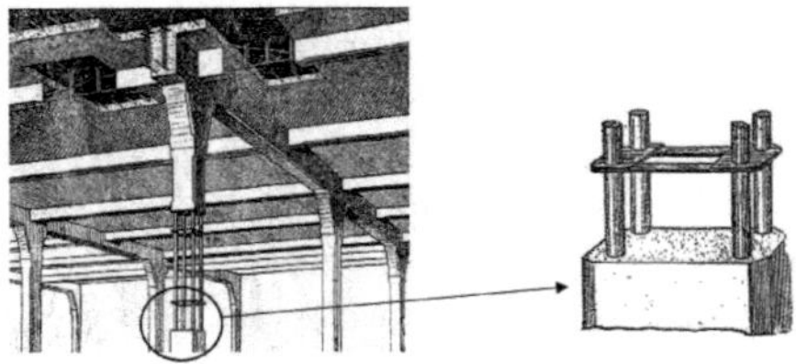

Source: Valenzuela, p.137

All this accumulation of new knowledge began to be applied following a rule or methodology, the theoretical model or design, tests of that model, physical tests, presence of errors, adjustments were made to improve the model and the same steps were followed again, This is called the scientific revolution from theory to technology, leaving a great legacy, where the theoretical science that was jealously guarded by a few is slowly disappearing giving way to applied science, a fusion of technique and practical science.

1.2.- Historical and Current Building Materials. Building materials are the most important elements in any constructive or structural form, it is the primary basis or its essence, these can be divided into several groups according to their composition and nature, Figure 1.7.

Figure No. 1.7

Construction Materials Groups

1.2.1.- Organic materials. These are those produced by nature of a vegetable nature, composed mainly of wood with all its derivatives and guadua.

- **Wood.** It is an excellent building material that man has used since ancient times, it is a renewable and sustainable resource that can be used as it is found in nature without making any changes, either physical or chemical. It is the oldest construction material, "one of the characteristics of wood that differentiates it from other structural materials is the fact that it is the only living material used" (Hernández, 2013, p. 39). This material is composed of elongated and hollow cells, whose axes run parallel to the length of the tree, similar to a set of thin-walled tubes attached to each other and their cells are agglomerated by means of a natural resin called lignin, (Keyser, 1982, p. 385). Timber trees develop in coniferous forest areas located in the cold and temperate zones of the northern hemisphere and to a lesser extent in the south, these forests are mainly composed of homogeneous coniferous species such as pines, firs, cypresses. In the tropical zones, broadleaf forests are found, composed of a great variety of timber species such as walnut, kapok, mahogany, chanul, cedar, etc. It is one of the healthiest building materials in existence, as it acts as a natural regulator of the interior environment, it is a living material that breathes and helps to maintain a healthy environment. ventilation, stabilises humidity, filters and purifies the air, (Ghoreishi, 2011, p. 29). Due to its good physical and mechanical properties, wood is widely used in the field of construction, either as a main structure for housing, in the field of ornamentation to build furniture, doors, windows, railings, wall and floor coverings, it is also widely used structurally in concrete buildings for the manufacture of formwork boards for beams, slabs, columns, reinforced walls. Another widely used type of timber is glulam, which is used for roof structures and structural beams to cover large spans. The use of wood as a construction material has the following characteristics, Figure No. 1.8.

Advantages: (i) Product of natural renewable origin

ii) Easy to work with and model, very versatile in use.

iii) Thermal insulation

iv) Durable when proper protective measures are taken

v) Good physical and mechanical properties

Disadvantages: (i) Highly sensitive to the environment exposed to weathering

ii) Highly combustible

iii) Vulnerable to external agents

iv) Limited dimensioning in its natural state

v) Indiscriminate felling of certain species in forests

Figure No. 1.8
Laminated wood structure

Source: Blog Maderera Andina, https://maderera-andina.com/maderera-andina-quieres-trabajar-con-madera- laminate-meets-some-projects/

- **Guadua.** Guadua is a woody plant belonging to the bamboo family, which in the world has approximately 1,250 species, of which in Colombia there are a large number of species of which the guadua angustifolia stands out, classified by the wise Humboldt with the scientific name of Bambusa guadua. This particular species grows abundantly in very fertile regions up to 1,700 metres above sea level, forming large extensions of guaduals, these climatic conditions make it ideal from an economic point of view for use in various fields, it is a sustainable and renewable resource that self-multiplies vegetatively, being a carbon dioxide retainer, (Hidalgo, 1978, p. 5- 6). In the construction industry it is widely used for enclosures, shoring, scaffolding, formwork, formwork, casetons; in rural areas it is an indispensable material in the construction of houses, Figure No. 1.9. The use of guadua as a construction material has the following characteristics:

Advantages: i) Due to its geometry and hollow interior, it is a lightweight material.

ii) The knots give it rigidity and elasticity.

iii) Highly versatile in housing construction

iv) As a pipe it is useful in the conduction of fluids.

Disadvantages: (i) Highly sensitive to the environment exposed to weathering

ii) Highly combustible

iii) Vulnerable to xylophagous agents if untreated

iv) Very varied dimensioning in its geometry

Figure No. 1.9
Guadual

Source: https://guaduabambucolombia.co/tag/tolima/

1.2.2.- Inorganic materials. These are the materials found in the lithosphere and which have been used by man since ancient times, mainly composed of minerals, among which are stone, earth, adobe and brick.

- **Stones.** Rocks are a set of aggregates composed of mineral particles with different dimensions and without a determined shape, this construction material is found forming large masses called deposits, which are exploited in the open air or quarries, it is the first material used primitive man. This material can be used as a resistant element when it is placed on the building site in its natural state, as a decorative element when it is worked beforehand to give it a specific shape, dimension or finish, and as a manufactured raw material when by-products are obtained which are subsequently used in other activities.

According to their constructive use, they can be used as foundations, cyclopean walls, dividing walls, floors, architectural and decorative finishes on façades and interiors of buildings, in roads and mainly in pavements, in vaults and arches, in dam embankments, among others, or as raw material to obtain other construction materials. The use of this as a construction material has the following characteristics, Figure No. 1.10.

Advantages: i) Very durable

ii) Acoustic and thermal insulation

iii) Resistant to atmospheric agents

iv) Physical and mechanical properties

Disadvantages: - i) Problems in transporting large rocks

ii) Very varied dimensioning in its geometry

iii) Longer execution time

iv) High manpower requirement

v) Materials with high absorption capacity

Figure No. 1.10
Stone construction

Source: http://castellonenarchivos.blogspot.com/p/construcciones-de-piedra-en-seco.html

• **Earth.** It is found in all places around the globe, it is one of the oldest materials used by man thousands of ago, its use in the construction of houses and walls was widely used, no great knowledge was needed since the technique of the trodden earth was used. The most primitive houses were built in the past with this material, today more than a third of the world's population lives in earthen houses, in places where it is traditional this type of construction is maintained, but in some developed countries, this material continues to be used in rural constructions, (Ghoreishi, 2011, p. 24). This construction method consisted of building a solid wall step by step using earth (mud) or wet clay mixed with sand and chopped straw, which were poured into a wooden formwork called tapial, to be compacted in situ with hand tampers in horizontal layers. Once this work is finished, the wall is left to dry and tapial is lifted or removed until the desired height of the wall is achieved, Figure No. 1.11. The use of trodden earth as a building material has the following characteristics:
Advantages: i) Good thermal, acoustic and fire-resistant insulation.

ii) Easy to obtain and environmentally friendly

iii) Highly versatile in self-build homes

iv) It is recyclable, its debris can be reintegrated into nature.

v) Soil is an inert, non-polluting and non-toxic material.

Disadvantages: i) Highly sensitive to the environment if exposed to the weather.

ii) Construction limited in height

iii) Vulnerable to water

iv) Seismic vulnerability

v) Carry out regular maintenance

Figure No. 1.11

Dwelling constructed with rammed earth or mud wall

Source: https://construyediferente.com/tapial-tecnica-antigua-nueva/

- **Adobe.** It is the first construction material created by man and still in use, due to its easy handmade elaboration that makes it self-buildable, with traditional craft techniques and its low economic cost. Its manufacturing process is carried out with clayey earth to which sand, water, chopped straw and cow dung or manure are added. All these materials are stirred together to form a manageable doughy mass that is placed in wooden geometric moulds, giving rise to the adobes in the earth, which are then left to dry covered places. Its use in construction is very varied, it can be used to build isolated vertical walls or interlaced to form dwellings, the union between the blocks is made with the same adobe material. The use of adobe as a construction material has the following characteristics, Figure No. 1.12.

Advantages: i) Good thermal, acoustic and fire-resistant insulation.

ii) Highly versatile in self-built homes

iii) Its debris can be reintegrated back into nature

Disadvantages: i) Highly sensitive to the environment if exposed to the weather.

ii) Construction limited in height

iii) Carry out regular maintenance

iv) Vulnerable to water

v) Seismic vulnerability

Figure No. 1.12

House built in adobe

Source: https://es.wikipedia.org/wiki/Adobe

- **Brick.** It is the first building manufactured by man through a production process, adding the firing to the old adobes, with this process is forming the first artificial stone by obtaining by this means physical and mechanical properties such as durability and strength, for these important characteristics the brick was gradually displacing the adobe. Its use in construction is very varied, it can be used to build vertical masonry walls, isolated or interlaced, to form high-rise houses, the union between the blocks is made with a mortar of lime and sand. The use of bricks as a construction material has the following characteristics, Figure No. 1.13.

Advantages: i) Good thermal, acoustic and fire-resistant insulation.

ii) No regular maintenance required

iii) Highly versatile in housing construction

iv) Your rubble can be reused

Disadvantages: (i) Energy-intensive in production

ii) Size variability and strength

iii) Vulnerable to the environment affecting durability

iv) Seismic vulnerability

Figure No. 1.13
Brick-built house

Fuente: https://ar.pinterest.com/pin/877709414870763298/

1.2.3.- Composite materials. These are materials that are not found in nature and are formed by the union of several raw materials, which are transformed by means of industrial processes to obtain other materials that are not found in nature, and are used by man in construction activities; these materials include lime, concrete, iron and its derivatives, and glass.

- **Lime.** Lime is the residue product of the calcination process at about 1,000 °C of limestone rocks called quicklime (calcium oxide), which is subsequently slaked with water to obtain a hydrated material in the form of paste or powder (calcium hydroxide). In antiquity, it was the Greeks who first used lime as a building material, in coatings or stuccoes for masonry walls, whether adobe or brick. Later it was used in the preparation of mortars, destined to join a series of masonry elements of stone, brick, to constitute a construction unit with its own characteristics, (Arredondo, 1989, p. 7- 8).

Lime mortars are responsible for the stability of the great constructions of ancient Greece, Rome and medieval constructions, but they were replaced with the invention of cement. Its use in construction is very varied, vertical masonry walls can be built in isolation or interlaced to form high-rise dwellings, the union between the blocks is made with a mortar of lime and sand, in the external plastering of masonry walls, mainly in the restoration of monuments and ancient architectural constructions. The use of lime as a building material has the following characteristics, Figure No. 1.14.
Advantages: i) It allows the buildings to breathe.

ii) Reabsorbs the carbon dioxide emitted by calcination

iii) Firing temperature lower than that of concrete

iv) It is breathable, hygroscopic and flame retardant.

v) Speed of construction with other elements

vi) Moisture does not remain in the walls

vii)It is aseptic, bactericidal and fungicidal.

viii) Good plasticity and workability

ix) Due to its volumetric stability there is no shrinkage.

x) Low risk of cracking
Disadvantages: - i) Low modulus of elasticity

ii) It is weak and decomposes faster than masonry.

iii) It sets slower than cement.

iv) Its structural use is very limited

Figure No. 1.14
House built in lime-plastered masonry.

Source:www.terram.cat/portfolio/arrebossat-de-terra-i-calc-sobre-mur-de-bales-de-palla-habitatge-https://unifamiliar-a-yelamos-de-arriba-guadalajara/?lang=en

• **Concrete.** For a long time limestone had been considered unsuitable for construction, nor for the manufacture of limes and mortars because lime contained clays, but in the 18th century in England it was found that some limes made with these limes with clay produced mortars that were more resistant than those made with pure limes and with a supremely important property: they set in the presence of water, which was not the case with the traditional mortars used up to that time (Arredondo, 1969, p. 7). In 1845 Isaac Johnson improved the process previously patented by Aspdin & Parker by obtaining a product called Clinker and called it Portland Cement, due to the similarity in colour of some limestone rocks from the

quarries on that island.
The new portland cement is a synthetic material (artificial rock) that sets and hardens by means of a chemical reaction with water, resulting from the controlled calcination at high temperatures of clayey and limestone materials, which are then mixed with water. clayey materials provide silica (SiO_2) and calcined limestones calcium silicate ($CaSiO_3$), which is why it is called hydraulic cement, (Keyser, 1982, p. 298). William Boutland in 1854 built the first reinforced concrete structure, which consisted of a two-storey house made of a mixture of concrete with iron and wire reinforcements, also Françoise Hennebique in 1892 made an important contribution by using cylindrical section bars, which could be bent and used as anchors (Lamuz, 2015, p. 14).The concrete obtained has the behaviour of an artificial rock, assimilating large compressive loads, but with a limitation to traction, but "it is necessary to compensate the destructive efforts by placing pieces of a material that is resistant to traction. This is the case of iron, which has long been used in masonry constructions, (Pérez, 2012, p. 7). The union of these two construction materials, called reinforced concrete, brings with it countless possibilities of constructive order and versatility of structural solutions, making the most of the physical and mechanical characteristics in the union of these materials, which work together in the works, being used in large construction and infrastructure works around the world.
Pérez concludes that "reinforced concrete allows the construction elements to be dissociated, the load-bearing wall disappears, all support elements are eliminated and the construction of corbels and cantilevers is permitted, thus expanding the catalogue of structural solutions", (2012, p. 10). Its use in construction is very varied, building structures can be constructed in reinforced concrete, vertical masonry walls isolated or interlocking to form high-rise housing, the union between the blocks is made with a mortar of cement and sand. The use of concrete as a building material has the following characteristics, Figure No. 1.15.

i) It is a universally accepted material.

ii) Easy availability of materials

iii) Adaptable to any structural and architectural form

iv) Physical and mechanical properties

v) Fire resistant

vi) High durability

vii)Requires very little maintenance

viii) Good plasticity and workability for construction

ix) Monolithism and continuity of structure

Disadvantages: i) High volume of construction and high costs.

ii) Structural seismic behaviour must be taken into account.

iii) Slow to set and set in place

iv) Non-structural elements are more gravity loads.

v) Requires large, heavy sections

vi) Shrinkage during the setting and hardening process

Figure No. 1.15

Construction of reinforced concrete building

Source: http://www.caminoymarchal.com/QuienesSomos.aspx

• **Iron and its derivatives.** The history of iron is linked to the development prehistoric mankind, its use became popular in some ancient societies of the Middle East, where metallurgical technology reached its peak around the twelfth century BC, called by historians the Iron Age, a period characterised mainly by the manufacture of agricultural tools, combat weapons. It is a metallic material, with magnetic properties, malleable and its colour is silvery white; it is a chemical element with symbol Fe, Atomic Number 26,located in Group 8 of the Periodic Table of the elements, it is the fourth most abundant element on Earth.

The production of reinforcing steels is carried out in steel plants for the construction industry, which consists of the alloying of iron with other metallic and non-metallic elements to obtain certain physical and mechanical properties. Steel is considered to be an iron alloy with a maximum carbon (C) content of 2% with the addition of small quantities of silicon (Si), manganese (Mn), phosphorus (P) and sulphur (S). Steel is one of the most versatile construction materials available, producing rebar, structural sections in all forms, structural angles, circular pipe, cold rolled flat sheet.

In the field of construction, they are used in metal buildings, in roofs, in foundations of structures, in road infrastructure such as metal bridges, their derivatives in hanging facades. The use of steel as a construction material has the following characteristics, Figure No. 1.16.

Advantages: (i) High tensile strength

ii) Good physical and mechanical properties

iii) Ease of joining several members with welding, rivets, bolts, pins, etc.
iv) Your rubble can be reused

v) Speed of construction with other elements

vi) Fatigue resistant

Disadvantages: - i) Maintenance costs

ii) Heat spreader

iii) Vulnerable to high temperatures

iv) Susceptibility to buckling if they are slender

v) Vulnerable to corrosion

Figure No. 1.16
Construction of metal bridge

Source:
http://www.adurcal.com/enlaces/mancomunidad/guia/tablate/puenteautooator.htm

- **Glass.** The history of glass dates back to approximately three thousand years B.C. in the fertile crescent, where the Phoenicians and Egyptians were the main manufacturers and suppliers of this material. With the expansion of the Roman Empire, many craftsmen were brought to Rome to teach their manufacturing and production techniques.

This material is obtained by mixing the raw materials of lime, sand, sodium carbonate and placing them in a fire at a temperature of 1500 °C, the amorphous structure of a hard and transparent material was obtained. It is a solid, hard, brittle, transparent material obtained by melting at high temperatures a mixture of sodium silicate (Na2SiO3), calcium (Ca), lead (Pb).

In the field of construction it is widely used as a decorative and architectural element, it is used in all types of housing, buildings, roofs, greenhouses, its derivatives such as flat glass, tempered safety glass, glass , in hanging facades, doors, windows of all types, shop windows in shopping centres, dividing walls. The use of glass as a building material has the following characteristics, Figure No. 1.17.

Advantages - i) High transparency or opacity

ii) Good physical and mechanical properties

iii) Thermal and acoustic insulation

iv) Your rubble can be reused

v) Speed of installation

vi) Weatherproof

Disadvantages: - i) Fragile to impact

ii) They do not tolerate sudden changes in temperature.

iii) Vulnerable to high temperatures

iv) Periodic maintenance

v) High cost

Figure No. 1.17

Building with floating glass façade

Source: https://vidsa.cr/fachadas-en-vidrio/

A summary of the properties (advantages and disadvantages) of the most commonly used building materials in this field is presented in Table 1.1.

Table No. 1.1

Summary of building material properties

TECHNIQUE CONSTRUCTIVE	ADVANTAGES	DISADVANTAGES
WOOD	Product of natural renewable origin, easy to work and shape, very versatile use, thermal insulator, durable when proper protective measures are taken, good physical and mechanical properties.	Very sensitive to the environment, exposed to weathering, highly combustible, vulnerable to external agents. external agents, limited size in their natural state, indiscriminate felling of certain species in forests.
GUADUA	Due to its geometry and hollow interior it is a light material, the knots give it rigidity and elasticity, very versatile in the construction of houses, as a pipe it is useful in the fluid conduction.	Highly sensitive to , highly combustible, vulnerable to xylophagous agents if untreated, dimensioning very varied in its geometry.
PÉTREOS	Very durable, acoustic and thermal insulation, resistant to atmospheric agents, good physical and mechanical properties.	Problems with transporting large rocks, very varied dimensioning in geometry, longer execution time, need for considerable manpower, materials with high absorption capacity.
	No drying shrinkage, low-wood constructions, good thermal insulation, fire resistant, good acoustic insulation, fire resistance, homogeneity of the wall, workmanship economical, no plastering required.	Highly sensitive to the environment when to weather, limited height construction, vulnerable to water, seismic vulnerability, perform regular maintenance
EARTH		

	Easy to obtain and environmentally friendly, very versatile in self-built housing, recyclable, its debris can be reintegrated into nature, it is an inert material, non-polluting and non-toxic.	
TECHNIQUE CONSTRUCTIVE	ADVANTAGES	DISADVANTAGES
ADOBE	Good thermal, acoustic and fire-resistant insulator, very versatile in self-built housing, its rubble can be reintegrated into the building site. nature, low energy consumption.	Very sensitive to the environment, exposed to weather, limited in height, vulnerable to water, vulnerability seismic, perform regular maintenance.
BRICK	Good thermal, acoustic and fire-resistant insulation, no need periodic maintenance, very versatile in housing construction, its rubble can be reused.	energy-intensive production, variability in size and strength, vulnerable to the environment affecting durability,seismic vulnerability.
CAL	It allows constructions to breathe, reabsorbs the carbon emitted by calcination, lower firing temperature than cement, it is breathable, hygroscopic and fireproof, rapid construction with other elements, moisture does not remain in the walls, it is aseptic, bactericidal and fungicidal, good plasticity and workability, due to its volumetric stability there is no shrinkage, under risk of cracking.	Low modulus of elasticity, weak and decomposes faster than masonry, slower setting than cement, very limited structural use.

CONCRETE	It is a universally accepted material, easy availability of constituent materials, adaptable to any structural and architectural form, good physical and mechanical properties, fire resistant, high durability, low maintenance, good plasticity and workability. For construction, monolithic and continuity of the structure.	High construction volume and high costs, structural seismic behaviour must be taken into account, slow setting and commissioning, non-structural elements are more gravity loads, requires large sections with high weight, contractions during the setting and hardening process, high quantities of non-renewable materials.
TECHNIQUE CONSTRUCTIVE	ADVANTAGES	DISADVANTAGES
IRON	High tensile strength, good physical and mechanical properties, easy to join several members with welding, rivets, bolts, their debris can be reused, fast construction with other elements, fatigue resistant.	maintenance costs, , vulnerable to high high temperatures, susceptibility to buckling if slender, vulnerable to corrosion.
GLASS	High transparency or opacity, good physical and mechanical properties, thermal and acoustic insulation, its debris can be reused, quick to install, resistant to corrosion and corrosion. weathering.	Fragile to impact, cannot withstand sudden temperature changes, vulnerable to high temperatures, periodic maintenance, high cost.

CHAPTER II

SUSTAINABLE CONSTRUCTION

UN World Commission on Environment and Development met in 1987, where several countries drafted the famous Brundtland Report, which resulted in the globally coined term Sustainable Development, which defines how humanity should behave in order to meet the needs of the present without compromising the opportunities for future generations to meet their own needs. Sustainable development is not an eminently ecological concept, it interacts in a balanced way in the triad of ecological, social and economic fields. Sustainable construction is a new building model, where environmental impacts related to the construction process of the building project are taken into account, "its main premises are to optimise the resources construction materials and regulate the consumption of these raw materials to the maximum", (Magaña, 2022, p. 76). Sustainability does not consist of keeping natural resources intact, but implies making efficient use of them, (Mata, 2009, p. 6).Sustainable construction presents different alternatives for the use of traditional materials, as well as ancestral construction techniques using adobe, rammed earth, mud, wattle and daub, wood and guadua in various rural and peasant constructions. Sustainable construction is based on the appropriate choice of materials, construction processes and energy saving, and also refers to the urban environment and its development (Casas, 2011, p. 12). Sustainable construction has made significant advances in recent decades, which has facilitated its use in the construction of more economical housing. Bedoya defines it as: "respectful and committed to the environment, makes a sustainable use of energy, minimises its impacts, reduces theenergy consumption, does not waste materials, but reuses and recycles...", (2011, p. 18),

Sustainable construction is based on the following premises, (Casas, 2011, p. 15-18):

- Adaptable and environmentally friendly
- Optimising the use of building materials.
- To regulate the consumption of raw materials as much as possible.
- Energy saving and use of renewable energies
- Use of typologies of construction techniques adapted to the area.
- It is stable, comfortable, safe, durable, functional, and secure.
- Controlling the generation of construction waste
- Correct integration with the environment, minimising the negative environmental impacts that may be generated during construction.
- Cost-effective construction.

Cruz emphasises that sustainable construction "reflects on the environmental impact of all the processes involved in a dwelling, from the procurement of manufacturing materials that do not produce toxic waste or much energy, and construction techniques that produce minimal environmental deterioration", (2013, p. 3).

It identifies the basic aspects of sustainable construction, (2013, p.

- Protection of the ecosystem on which it sits
- Conservation of material and water resources
- Energy systems that promote energy savings
- Building materials and their life cycle
- Recycling and reuse of waste
- Proper control of construction waste
- Integration of alternative energy sources
- Efficient designs for minimum energy consumption

2.1.- Sustainable Building Materials. Sustainable building materials that should be used in the construction of sustainable housing are those that have a low environmental impact, are durable, require little maintenance and can be recycled, reused or recovered. The criteria for choosing these materials are as follows, (Ghoreishi; 2011, p. 12):

- Health, materials must be natural and free of toxins.
- Ecological, they must be locally sourced, with low impact at the time of extraction and transport.
- Ethical, that have a social recovery in their production by promoting productive activities and trades.
- Sustainability, that materials have a low environmental impact and are sustainable in their life cycle.
- Recycling and reuse, if they meet these characteristics.
- Toxic emissions, must be free of toxic emissions.
- Energy criterion, it must produce the lowest energy in its production.

In the field of construction, the selection of sustainable materials has become a very important factor to consider, mainly because of their manufacture and other inherent factors such as the environmental footprint, which have an impact on the environment. The following is a list of sustainable materials that have a low impact on the environment:

- ❖ Recovered and recycled sawn timber
- ❖ Recycled steel
- ❖ Natural clay or baked clay
- ❖ Recycled plastic and rubber
- ❖ Natural stone

- ❖ Cellulose
- ❖ Bamboo
- ❖ Thermochronic cement
- ❖ Self-repairing concrete
- ❖ Recycled synthetic shingles
- ❖ 3D printed materials

2.2.- Life Cycle of Construction Materials. All construction materials production processes have a life cycle, which is an effective tool that provides objective information about the materials and construction processes. Life Cycle Assessment (LCA) is a methodology used to measure the environmental impact of a project or product throughout its service life, by establishing the initial analysis for the reduction of carbon emissions from any building or production process. The life cycle (LCA) is considered to have four stages, which are listed and configured in Figure No. 2.1, (MADS, 2022, p. 36):

- ► Stage I, Manufacturing, (Extraction of raw materials, manufacturing process)
- ► Stage II, Construction, (Distribution and transport of raw materials, commissioning and construction)
- ► Stage III, Use and Maintenance, (Lifetime of the building)
- ► Stage IV, Deconstruction, (, waste disposal, reuse, recovery, recycling)

Figure No. 2.1

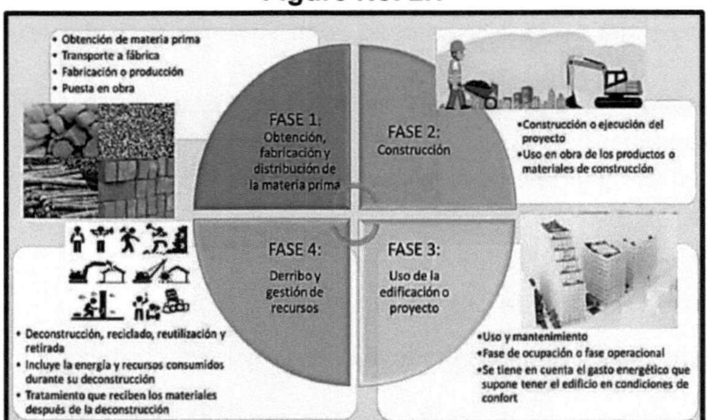

Life cycle of building materials

Source:https://www.minambiente.gov.co/wp-content/uploads/2023/06/Guia-de-materiales-para-la-construccion- sustainable.pdf

2.3.- The Circular Economy in the Construction Sector. The Circular Economy (CE) is a methodology and model that represents a fundamental change in production and consumption, it has the following objectives, Figure No. 2.2, (MADS, 2022, p. 15):

- It considers environmental impacts throughout the life cycle of a product and is integrated from its conception.

- Reintroduce products that have completed their life cycle back into the economic circuit.
- Reuse certain products that can still work in the production new recycled products.
- Implement a second cycle for deficient and imperfect materials.
- Recycle certain products that are present in construction waste.

Figure No. 2.2

Circular economy of the construction sector

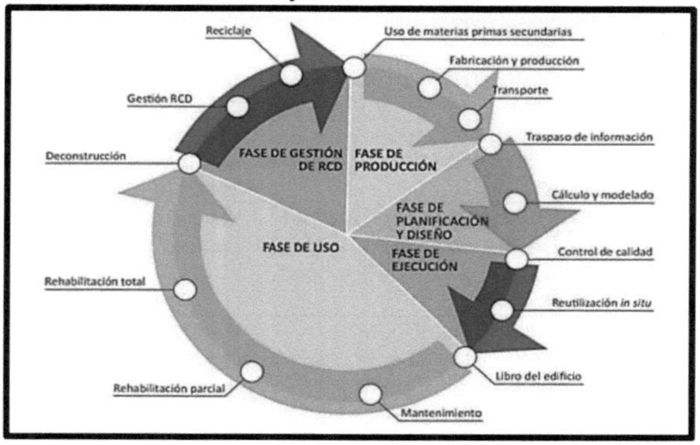

Source: https://www.minambiente.gov.co/wp-content/uploads/2023/06/Guia-de-materiales-para-la-construccion- sustainable.pdf

2.3.1.- Main construction materials. The construction materials come from raw materials or processed products, they are the components of the different construction elements of a structure or a product. building. There is a very wide variety of materials used in the construction sector, which are mainly used in all kinds of projects, such as buildings in general and various infrastructure constructions. The most commonly used building materials are presented and classified, taking into account their function in the field of construction and the raw materials used in their industrial process, Table No. 2.1, (MADS, 2022, p. 34-35):

Table No. 2.1

Group of materials used in construction

MATERIAL	FIELD OF CONSTRUCTION	RAW MATERIALS
Steel	Wires, steel structures of beams, columns, slabs, floors, etc.	Ferrous materials, coal, chemical materials, iron scrap
Asphalt	Track pavements	Stone aggregates, asphalt emulsion or bitumen
Cement	Concretes in general, mortars as wall coverings, walls and floors, tiles	Limestone rock, clay soil, gypsum ore, water
Ceramic	Masonry walls, roof tiles, floor coverings, wall coverings	Clays, sands, kaolin, gypsum, silica, feldspar, talc, water
Drywall sheets	Dry construction, wall and ceiling cladding in interiors	plasters, sheets of cardboard or paper, glass fibres
Superboard sheets	Dry construction, wall and ceiling cladding interior and exterior false ceilings	Cement, cellulose fibres, quartz, water
Concrete	Structures of beams, columns, slabs, floors, walls, walls, walls	Cement, stone aggregates, water
Painting	Surface coatings of walls, ceilings, structures, etc. metallic	Lime, gypsum, kaolin, aluminium, titanium dioxide, water, mineral solvents, resins
Wood	One- and two-storey dwellings, structural formwork panels, general utensils, furniture	Natural trees of different varieties
Stone materials	Concretes, coating mortars	Mineral rocks from quarries, mineral aggregates from rivers,

2.4.- Environmental Problems of Construction Materials. The daily generation of rubble or construction and demolition waste (CDW), a product of the engineering activities of the construction companies and of the construction and demolition of buildings, is the main cause of the environmental problem of construction materials. The waste is not biodegradable and therefore has an environmental impact for any city or town, these materials are not biodegradable. Due to the lack of planning for an adequate management of these materials, municipal governmental entities, in order to provide a partial solution to this problem, require large and costly extensions of land for their final disposal. The sites suitable for this purpose are generally located on the outskirts of the city, in order to avoid as much as possible the production of pollution and agents harmful to the health of the citizens living in the surrounding areas. The occupation of unauthorised areas for the direct dumping of tons of construction waste, without any kind of treatment and in an uncontrolled manner, causes environmental problems, mainly in the degradation of the landscape and the environment, Figure No. 2.3. Taking into account the problem generated, it is necessary to address its solution by thinking about the development of new clean technologies, with the aim of initiating processes that involve adequate management in the use of these wastes, allowing the maximum use of all these materials, representing at the same time a benefit for society and thus reducing the negative impact that these wastes produce on the environment (Magaña, 2022, p. 72).

Figure No. 2.3
Disposal of construction debris

Construction debris or solid waste cannot be treated as household waste, as a high percentage of it can be reused if it is subjected to a recycling process. This process consists of the use and reuse of solid waste applied in construction activities, particularly in the production of prefabricated elements such as masonry blocks, bricks, paving stones, aggregates for construction, which would be an alternative to solve a large part of this environmental problem. By optimising the recycling of these products, new poles of economic development would be created, promoting new industries and countless jobs and underemployment under the premise of recycling, reuse and valorisation of these ecological materials.

2.4.1.- Typology of construction rubble. The importance of the construction sector in economies is very significant. Currently, this industry is undergoing a period of unprecedented growth and expansion worldwide, which implies a large daily consumption of energy and non-renewable materials, reaching an average value of 60% of the materials extracted from the subsoil. The percentage distribution in volume of the consumption of raw materials used in the construction field is in Figure 2.4, (Magaña, 2022, p. 73).

Figure No. 2.4

Consumption of materials in construction

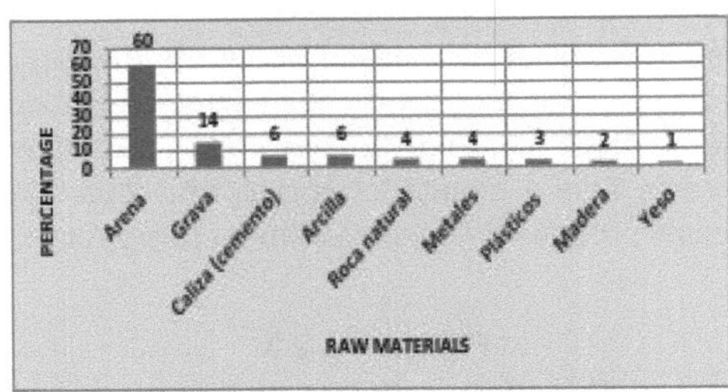

This construction activity generates a high volume of waste called construction debris (CDW), which is not classified as Urban Solid Waste from the cities (also called waste).domestic). Construction debris has different origins, depending on the generating activity, it is produced in different ways, one of them takes place in the construction of buildings, residential houses, public works of all kinds, where construction debris is the product of the handling on site by the technical staff, of raw materials and materials that are in optimal conditions, which for reasons of handling, placement and cutting, suffer a process of partial or total degradation that makes them unusable in the work. The other way in which construction debris produced is mainly due to total or partial renovations and demolitions carried out on a construction site that has been under construction or in permanent service for several years. These materials were in optimal conditions of quality and service, which, for reasons of disposal and handling, have subjected to an irreversible process of degradation. The main sources of production, the types of debris and the possible material to be recovered are identified in Table 2.2 below.

Table No. 2.2

Typology of construction debris

TYPES OF WASTE	SOURCE OF DEBRIS	MATERIALS RECYCLE	PERCENTAGE (%)
Demolitions: masonry, cyclopean and reinforced concrete, veneer, wood, glass, plastics, simple concrete floors, tile floors, tiles, cement and stoneware pipes.	One- and two-storey residential buildings, industrial buildings. Buildings from multi-storey buildings, civil constructions. Laboratories for Aggregates and pavements.	Bricks, roof tiles, tiles, concretes, screeds, mortars, mortars	10
Constructions: brick cuttings, fractured blocks, broken tiles, remains of mortar and concrete mixes, tile cuttings, façade tiles, floor tiles, etc. fractured.	Residential or industrial or industrial buildings, single or multi-storey.	All material can be used	20
Repair, Rehabilitation, Maintenance: same materials as for demolition works.	Multi-storey buildings, one- or two-storey residences, buildings industrial, civil works.	Bricks, roof tiles, tiles, concretes, screeds, mortars, mortars	70

Source: Adapted from https://www.concretonline.com/rcd-demolicion/reciclaje-y-reutilizacion-de-materiales-residuales-de- construccion-y-demolicion

2.4.2.- Composition of construction debris. The composition of construction debris depends mainly on the activities generating the debris, the characteristics of the works and the construction practices employed. In Table 2.1, the construction activities of repair, rehabilitation and maintenance are the main sources of debris production with 70%. Figure 2.5 the results and the percentage of materials produced in these activities, where ceramic materials and concrete make up 66% of these (Gaiker, 2007, p. 55).

Figure No. 2.5

Composition of construction debris

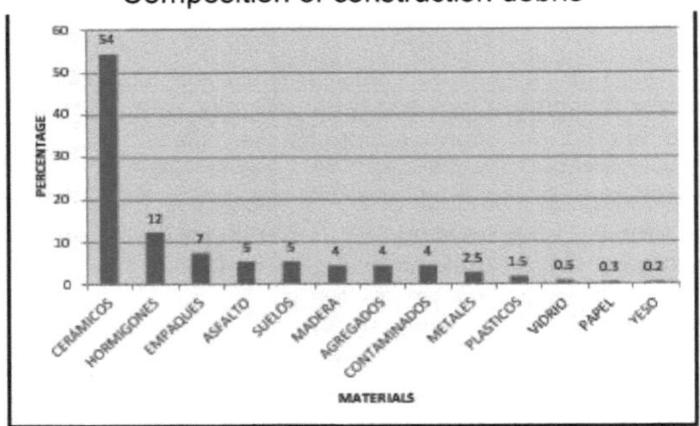

2.4.3.- Classification of construction debris. According to the European Waste List (EWL), construction debris is classified into three categories, A) Hazardous Waste, B) Non-Hazardous Waste, C) Inert Non-Hazardous Waste, (Gaiker, 2007, p. 52-54).

A) **HAZARDOUS WASTE.** These are construction materials which, when handled, contain high proportions of materials that are harmful to the environment. health and the environment, either through direct or indirect exposure, Table No. 2.3.

Table No. 2.3

Hazardous construction waste

RESIDUES DE CONTRUCTION (RCD) PELIGROSES	TYPES OF WASTE
	waste adhesives, glues, paints and varnishes, containing organic solvents or other dangerous substances
	Engine oils, transmission oils and lubricants
	Solvents
	Contaminated plastic and metal packaging
	Oil filters
	Lead batteries
	Coal tar and tarry products
	Metal waste contaminated with hazardous substances
	Soil and stones containing hazardous substances
	Insulation and construction materials, containing asbestos
	Construction materials containing gypsum, contaminated with dangerous substances
	Construction and demolition wastes containing mercury
	polychlorinated biphenyls (PCB) containing construction wastes
	Fluorescent tubes

B) **NON-HAZARDOUS WASTE.** These are construction materials, which by their nature can be treated or stored on site, Table No. 2.4.

Table No. 2.4

Non-hazardous construction waste

	TYPES OF WASTE
RESIDUESDE CONSTRUCTION (RCD) APPROVECHABLES	Wood
	Copper, bronze, brass
	Aluminium
	Lead
	Zinc
	Iron and steel
	Tin
	Mixed metals
	Insulation materials, not containing asbestos
	Gypsum building materials, uncontaminated

C) **INERT NON-HAZARDOUS WASTE.** These are materials that do not represent any kind of health risk when handled and stored on site, Table 2.5.

Table No. 2.5

Inert non-hazardous construction waste

	TYPES OF WASTE
RESIDUESINNE RTS	Concrete
	Bricks
	Roof tiles and ceramic materials
	Mixtures of concrete, bricks, tiles and ceramic materials
	Glass

For all of the above reasons, it is necessary to search for new construction alternatives that are environmentally friendly, mainly for raw materials for the production of recycled ecological materials, with aim of proposing technological solutions that optimise the use of these available resources, which results in the well-being of community and the improvement of the environment.

2.4.4.- Energy consumption and CO_2 emissions of construction materials. The construction sector or industry is one of the main factors in the economic growth of any country, but at the same time it is one of the biggest consumers of non-renewable materials. Historically, its procurement has undergone a quite remarkable change, as its extraction has been efficiently industrialised, its production and consumption has increased, becoming a highly impacting activity, due to its high energy consumption and the incidence of CO2 emissions due to this sector.

The environmental pollution produced by existing buildings and according to their lifespan, are a direct cause of environmental pollution by producing uncontrolled emissions of greenhouse gases, due to the high consumption of energy and water, as well as the different raw materials used in their operation. In order to understand the environmental impacts produced by the construction industry and its main materials, Table 2.6 shows the energy consumption and CO2 production for each of the main construction materials (Salazar, 2012, p. 7).

Table No. 2.6

Energy consumption and CO2 emissions of building materials

MATERIAL	CONSUMPTION ENERGY (MJ/T)	CO2 EMISSION (T/T)
Steel	11083	2,705
Coarse aggregates	177,2	0,010
Fine aggregates	494,6	0,021
River sand	121,7	0,010
Paving base	324,2	0,013
Cal	7670	0,798
Cement	7506	1,096
Ceramics	1172	0,830
Copper	98391	8,622
Guadua	1334	0,107
Brick - clay tiles	2750	0,243
Superboard sheets	8863	0,052
Wood	500	0,000
Painting	5247	0,408
PVC	72276	7,659
Flat glass	28952	1,859
Plaster	1190	0,205

From all of the above it can be concluded that the construction sector, despite being one of the main drivers of the global economy, efforts must be made to move towards a construction model that in the short term saves energy, that is consistent with natural resources and particularly with the proper management of construction and demolition waste (CDW), with the aim that this productive sector really becomes a model of sustainable construction, which is adapted and respectful of its environment, through the use of materials with low environmental and social impact throughout the life cycle of its buildings.Table 2.7 presents the environmental impact that is affected, in the most important sectors of some building materials, (ISTAS, 2005, p. 34).

Table No. 2.7

Environmental impact of the main construction materials

	MAIN ENVIRONMENTAL IMPACTS						
MATERIAL	GREEN HOUSE EFFECT	MARINE ACIDIFICATION	ATMOSPHERIC POLLUTION	OZONE LAYER	HEAVY METALS	ENERGY	CDW WASTE
Steel	++	++	+++	+	++	++	+
Aluminium	+++	+++	++	+	+++	+++	+
Ceramics	+	+	+	+	+	+	+++
Timbers	+	+	+	+	+	+	+
Stony	+	+	+	+	+	+	+++
Polystyrene	++	+++	+++	++	+++	+++	++
Polyurethane	+++	++	+++	+++	++	++	+
PVC	++	++	+++	+	++	++	++

NOTE:+ (Low impact) -++ (Medium impact) -+++ (High impact). Adapted from ISTAS, p. 34.

2.4.5.- Recyclable building materials. New construction using recycled building materials is the innovative construction process through the reuse of disposable materials, which is gaining importance in this field. From a sustainable point of view, the main recycled materials for the construction industry are:

► Metals. Mainly iron, steel and aluminium are part of these recycled materials, as they represent an important saving. The great advantage is that these materials can be recycled several times to be used again for the production of structural elements for construction.

► Stony. It is composed of waste concrete, gravels and sands, ceramic brick blocks, asphalt pavements, leftover mortar and concrete mixtures, among others. Despite having a low environmental impact, its main feature is its high durability, due to its treatment and proper use of construction waste, recycled aggregates are produced, which can be used in various construction activities, such as adaptation and filling of land, even mixed with cement and water. low-strength structural elements can be built, taking into account their appropriate technical design.

► Wood. This kind of materials can be considered as one of the most important ones. sustainable in the field of construction, due to this characteristic, they are recycled as materials for building supports and formwork, for the manufacture of chipboard, including for the manufacture of other non-constructive elements and as biomass in their energy assessment.

► Expanded polystyrene. It is a material that can be recycled and converted into in raw material for the new production of insulating elements, in walls, ceilings, roofs.

CHAPTER III.

PRINCIPLES OF TENDON STRUCTURES

The Spanish engineer-architect Felix Cardellach in his work "Filosofía de las Estructuras", explains that, for the builders of that time, the greatest mathematical discovery made lies in the laws of action and reaction of matter or constructive structures, before the acting external forces, baptised as the philosopher's stone of the Resistance of Materials (mechanical-constructive), (1910, p. 9-11). He defines that "the beings of all natural kingdoms, because they are exposed to the laws of external forces (gravity, wind, etc.) satisfy a general mechanical principle, without which their stability and resistance would not be possible, and this principle is none other than that of structure", (1910, p. 15).He also explains that man must be contemplative of nature, which is his great truth of the whole material world around him, helped by his imagination he argues, reasons and expresses phenomena: "in the form of geometrical concepts and analytical mathematical expressions, constituting a spiritual gear of rational deductions which lead us swiftly to the discovery of new truths", (1910, p. 15). All human beings and especially the constructions with their structures are subject to the laws of the external forces of nature, calling it the general mechanical principle, which must be satisfied by simplifying the complexity of natural phenomena by the conventional establishment of simple hypotheses, with the aim of being possible the stability and resistance of the structure, to this he calls the principle of the structure, where in nature we have a masterly and very representative example of this which is the skeleton of an animal, (1910, p. 15-18), concluding this way, that is to say, the skeleton of an animal, (1910, p. 15-18). 15-18), thus concluding that the engineer with this innovative spirit must discover new forms of resistant frameworks for the current construction.The sole origin of these structural and constructive forms lies in a higher level mechanical sensibility and natural inspiration, which are innate in man (architect, engineer) and which have constituted an important characteristic of his development (1910, p. 21). Cardellach summarises that: "the structural forms of construction should be classified or grouped by harmony of texture, mechanics and construction into two main groups:

1. Two-strand forms, which are able to withstand compressive and tensile stresses, use steel tendon construction systems in the form of a lattice (skeleton) and articulated with each other to absorb these stresses.
2. Uni-bearing forms, which are suitable for supporting only compressive stresses or only tensile stresses, are used as a construction system of segments composed of flexible elements", (1910, p. 25- 26).

He states that the law of the structural principle, which is dominant in construction by means of material diagrams of active lines, and expresses the structural principle of a construction, has now been discovered (1910, p. 18). From the anatomical or physiological point of view of all constructions made by builders, "it allows us to discover further ramifications that contribute to the harmony and method for their study...within the pseudo-elastic forms, we find two different species, structures without tendon and tendon-like structures", (1910, p. 27). The structural configurations in the construction forms of organic principle, initially pointed out to the metallic tendon its appropriate place in the masonry and due to economic circumstances facilitated the development of the tendon system, where the masonry passes to the background being protective of the lattice system by fixing the tensioned tendon and transmitting its strength to the concrete, imitating particularly the zoological architecture of vertebrates, with its composition of the skeleton, muscles and tendons, (1910, p. 101). This strategic placement of the tendon system is a way of solving the structural problem, where this branching reaches the appearance of a network in the very core of the structure and in connection with it, is achieved by the special construction method used, which consists of shaping the structure. Cardellach, concludes that: "the finesse and elegance achieved by the tendinous system in the astonishing resolution of a problem of structural resistance, obtaining an original and economical structure, resistant and elastic, which synthesises in an admirable way the advantages of construction that are characteristic of reinforced concrete", (1910, p. 126-128), putting an end to the mechanical rationalisation of the structures built by the Romans, as a consequence of the general characteristics of concrete. "With such factors alone, the man of ingenuity will always find in the vast field of his imagination all the structural forms urgently required by modern life", (1910, p.300).

In 1957, Torroja states that: "reinforced concrete is an organically constituted stone, within whose mass the tendon complex of the reinforcement is optimally distributed, it is dosed to provide the concrete with the tensile strength it needs at each point, the joint work between the mass of concrete and the steel bars is confined to adhesion, ensuring the transmission of stresses from the reinforcement to the concrete and vice versa", (2010, p. 67-68).From the point of view of mechanical properties, concrete is a different material to reinforcing steel, as it is very weak when working in tension compared to its good performance in compression, for this reason in the structures, reinforcing rod lattices are placed in the regions where tensile stresses occur, resulting in the cross section of the structure behaving as a section composed of two materials, concrete and reinforcing steel, (White et al., 1980). Fernández adds, taking up Cardellach's thoughts, that: "tendon systems are constructive forms that have their origin in the zoological architecture of vertebrates, such as the combination of skeleton, muscles and tendons", (2016, p. 108), since there are no structural forms previously established by nature, but these arise from basic principles. In the tendon system, the tendon acts as an element of the tendon system. The tendon is a tensile strength tendon, strategically located in the most vulnerable regions of the structure. Structurally, it defines three classes of tendons:

1. Tendons by ligament, used to join or bind critical points in masonry stone blocks in order to prevent partial slippage and maintain the monolithism of the structure.

2. Apparent tendons, used to bind the entire contact surface of the elements together in order to improve the mechanical strength of all the elements.
3. Bonded tendons, used to be placed internally in the structure to form a single monolithic structure.
The construction technology of bonded tendons later gave rise to reinforced concrete (Figure No. 3.1), where the internally placed tendon system resembles the work of steel bars, assimilating the mechanical tensile-compressive stresses and thus preserving the monolithism of the structure. For Cardellach, reinforced concrete was the most perfect technological material available to builders up to that time (2016, p. 108-109).

3.1.- Construction Systems Used in Colombia. The construction systems make up the set of methods, techniques and procedures used in the field of construction, which include the materials used, the equipment and tools used, which are rationally combined from a constructive and professional point of view, in order to generate and obtain a building.

Figure No. 3.1

Iron reinforcements in stone, St Genevieve Church in Paris

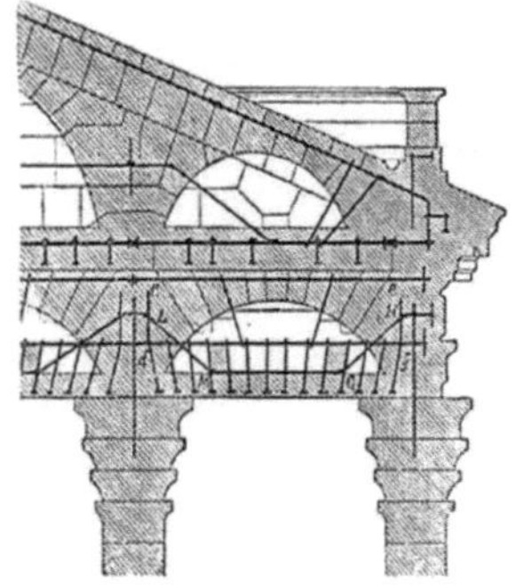

Source: Fernández (2016, p. 112)

The most common construction systems in our country are differentiated by the structural behaviour of their elements when certain structural stresses occur. These systems are classified into the following categories, which are summarised in Figure No. 3.2:

❖ Confined masonry, consisting of the traditional wall system of The brickwork is made up of reinforced concrete elements, which account for 62% of the total.

❖ Industrialised construction systems, where they are manufactured in a factory The proportion of all the structural elements in an automated way, in order to subsequently place these prefabricated elements on site, is 19%.

❖ Structural masonry, where the construction walls have a specific structural function, account for 15%.

❖ Other construction systems, representing 4%.

Figure No. 3.2

Building systems in Colombia

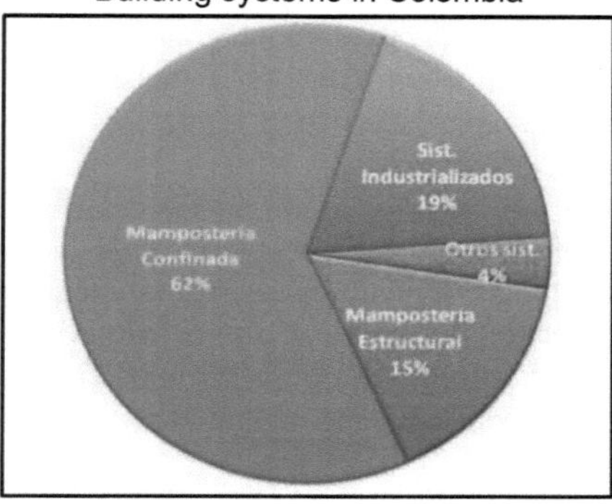

Source: (Salazar, 2012, p. 4)

The confined masonry construction systems use 99.5% natural materials, the structural masonry system uses 97.6% of natural materials, and these systems are the largest consumers of non-renewable natural materials, while the industrialised system uses 90.0% and is the smallest consumer of these materials, Figure No. 3.3.

Figure No. 3.3

Consumption of construction materials in Colombia

TABLA DE RESUMEN CONSUMO DE MATERIALES kg/m² y DISTRIBUCIÓN EN PORCENTAJE

CONSOLIDADO DE MATERIALES SISTEMA INDUSTRIALIZADO TOTAL	DISTRIBUCIÓN POR TIPO DE SISTEMA						
	kg/m²				Distribución %		
	SISTEMA INDUSTRIALIZADO	MAMPOSTERÍA ESTRUCTURAL	MAMPOSTERÍA CONFINADA	GUADUA - TIERRA ESTABILIZADA	SISTEMA INDUSTRIALIZADO	MAMPOSTERÍA ESTRUCTURAL	MAMPOSTERÍA CONFINADA
AGREGADOS TRITURADOS	536,5	399,2	625,0	90,3	42,44%	28,28%	25,96%
ARENA DE RIO	440,9	356,5	733,6	64,2	34,87%	25,25%	30,48%
CEMENTO GRIS	160,9	138,8	306,1	28,0	12,73%	9,83%	12,72%
ROCA MUERTA - TIERRA EXCAVACIÓN	40,6	162,4	372,5	841,7	3,21%	11,51%	15,47%
CERÁMICA COCIDA	43,9	320,8	358,1	4,3	3,47%	22,73%	14,87%
ACERO	29,5	21,0	9,4	2,2	2,33%	1,49%	0,39%
MADERA	5,4	3,3	0,1	105,5	0,43%	0,24%	0,01%
TEJA FIBROCEMENTO	3,1	6,4	0,0	0,0	0,25%	0,45%	0,00%
OTROS (PVC,COBRE,CEMENTO BLANCO,PINTURAS)	3,4	3,3	2,4	97,3	0,27%	0,23%	0,10%
TOTALES	1.264,3	1.411,7	2.407,3	1.233,5	100,00%	100,00%	100,00%

Source: (Salazar, 2012, p. 6)

3.2.- Non-Conventional Construction Systems. The development and modification of traditional construction techniques require evolution and adaptation to the needs of societies, which are directly related to safeguarding the integrity of their inhabitants and the right to decent housing; however, we must consider the increase in climate changes such as floods, the increase in extreme temperatures and droughts that have occurred in the last 10 years (Management Solutions, 2020). As a result of the above, it is necessary to design and propose new construction techniques that meet the structural requirements to reduce the high housing deficit in Colombia. In addition to meeting the challenge of structural requirements, it is necessary that these proposals reduce the construction time of a building and maintain optimum performance between materials, labour and equipment with proper planning and execution of its construction elements.

The need for a paradigm shift with regard to traditional construction methods and the handling of materials from a technological point of view in order to implement and build more economical and efficient environmental housing, the non-traditional construction system of tendon walls, is not a new methodology, but offers a viable and sustainable alternative through the analysis and chronological account of the technical and constructive evolution of the tendon wall.

3.2.1.- Alternative construction of Tendinous Walls. This new non-conventional construction system is called tendinous wall, and was developed in the 90's by the architects of the Universidad del Valle (Colombia) Supelano and Thomas, who present the Tendinous System as "an investigation that integrates design, technology and culture, which responds to the felt need to live in a house made of material" (2002, p. 25), it is a non-conventional construction alternative that is based on traditional architectural knowledge and techniques. 25), it is a non-conventional construction alternative based on traditional architectural knowledge and techniques. The aim is to optimise this knowledge and respond to the need of these communities to build their own homes using more resistant, safe and durable construction materials (2015, p. 170). This construction method has been applied in the department of Valle del Cauca in an efficient manner, with very good results in the execution of one and two-storey houses in different geographical locations, due to the ease and speed of its execution by self-builders. In addition to these constructive characteristics, it should be added its light weight of the structure and particularly its low economic cost, giving in this way an answer to the need for housing in rural and urban areas, of populations with a number of inhabitants of less than 50,000 inhabitants, representing for these communities many options of social, economic and particularly environmental character.

3.3.- Characterisation of Tendon Walls. Tendon walls represent a novel non-conventional construction system, its etymology is taken from Cardellach's writings on tendon structures; Supelano and Thomas take it up again by using the word "tendon", when a metallic wire is used as a structural element and on top of this a "tendon" of a natural mesh is placed, which when these two terms are united gives as a result:

TENDON+ TENDON= TENDON TENDON

The typology of this construction system consists of the on-site manufacture of flat modular panels, joined together by beams and columns or pillars of natural and/or metallic materials, with mortar coating, forming a structural confinement as the main component for the support of these constructions, taking into account a technical modulation of the panels. The constructive facility of this construction system of tendon walls for one- and two-storey houses has been successfully used in several cities in western Colombia, as it represents many environmental, economic and social options. In the area of the coffee-growing region of Valle del Cauca, with the help of the Federación Nacional de Cafeteros de Colombia in the decade of the 90's, implemented this new construction system, where approximately six hundred (600) houses of VIS character were built in the following urban areas of the municipalities:

- Bolívar, urbanisation Los Laureles
- Caicedonia, urbanisation Samaria
- La Victoria, Buenaventura housing estate
- Restrepo, urbanisation La Independencia
- Sevilla, urbanisation Villa Paz
- Trujillo, urbanisation La Paz

3.3.1.- Components of Tendon Walls. Tendinous walls are made in situ by forming flat modular panels, reinforced inside with a barbed wire mesh that serves as integrated tendons, joined by means of staples to vertical and horizontal elements forming a rigid structural frame composed of sawn or round wood, guadua, metal angle iron or reinforced concrete. The core of the partitions is a biodegradable natural fibre lattice called cabuya, mezcal or fique, which is intertwined with the barbed wire lattice, in to serve as a support for the application of the mortar mixture in successive layers to provide union between its elements and give rigidity to the lattice, constituting a monolithic structural element that will behave as a confined wall, giving a final finish to the panel with an average thickness of 4 to 5 centimetres (Torres, 2013, p. 1). 1). The basic construction materials in the composition of a tendon wall are (Figure No. 3.4 and Figure No. 3.5):

- Sinewy wall in roundwood or sawnwood

• Vertical and horizontal pars for the wall frame
• Iron nails for fixing the joints of the timber frame
• Barbed wire as fixing tendon
• Sack of fique intertwined with barbed wire
• Portland cement for mortar mix and composition

- Sand as an aggregate for mortar mix and composition
- Mixing water and mortar composition

Figure No. 3.4

Tendon wall system materials

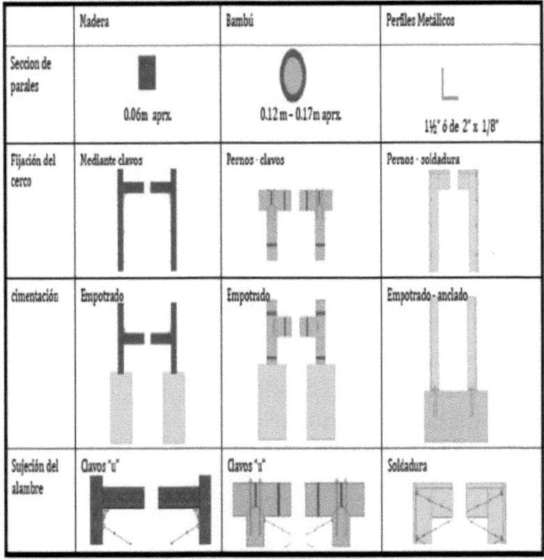

	Madera	Bambú	Perfiles Metálicos
Seccion de parales	0.06m aprx.	0.12 m - 0.17m aprx.	1½" ó de 2" x 1/8"
Fijación del cerco	Mediante clavos	Pernos - clavos	Pernos - soldadura
cimentación	Empotrado	Empotrado	Empotrado - anclado
Sujeción del alambre	Clavos "u"	Clavos "u"	Soldadura

Note: Taken from https://dokumen.tips/documents/3-muros-tendinosos-1.html -

Figure No. 3.5

Basic materials of the tendon wall system

https://www.maderastecnicasinmunizadas.co/producto/madera-aserrada-pino-y-eucalipto-inmunizado/ - https://www.facebook.com/media/set/?set=a.1132498343437065&type=3&_rdr - https://www.bambukindus.com/service/1-guadua-rolliza-angustifolia-kunth-tipo-macana/ - https://www.siderperu.com.pe/sites/pe_gerdau/files/PDF/FT%20ANGULOS%20DOBLADOS%20SP%2022mar19.pdf

Tendon wall in bamboo or bamboo

- Vertical and horizontal pars for the wall frame
- Metal bolts or iron nails for fixing the joints of the timber frame
- Barbed wire as fixing tendon
- Sack of fique intertwined with barbed wire
- Portland cement for mortar mix and composition
- Sand as an aggregate for mortar mix and composition
- Mixing water and mortar composition

► Tendon wall in metal profiles

- Vertical and horizontal pars for the wall frame
- Iron pins or welding electrodes, for fixing the joints of the timber frame
- Barbed wire as fixing tendon
- Sack of fique intertwined with barbed wire
- Portland cement for mortar mix and composition
- Sand as an aggregate for mortar mix and composition
- Mixing water and mortar composition

The tendon wall is composed of the following basic construction elements, in the constructive and structural formation (Figure No. 3.6):

- Gantry frame and geometric structural framing
- Flat modular partition wall composed of barbed wire tendons, the wire interlaced with core of sacks or sacking made of fique
- Mortar mix bonded to the fique core
- Component of the tendon wall embedded in the foundation structure

Figure No. 3.6

Components of the tendon wall system

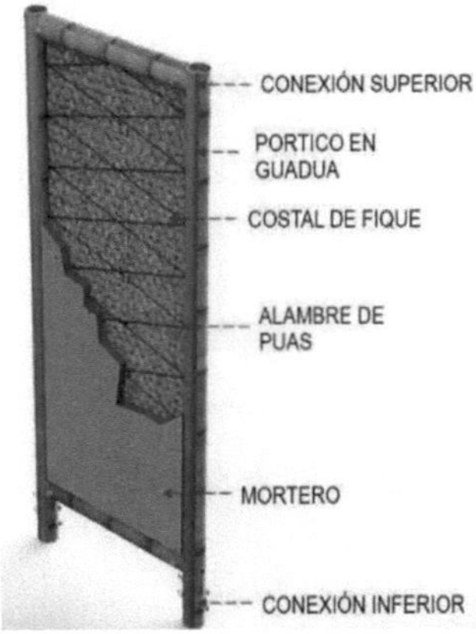

Note. Taken from: Mora (2022, p. 47).

3.3.2.- Construction of Tendon Walls. For the construction of a tendon wall, some geometric rules or standards must be met, as can be seen in Figure No. 3.7, (Velásquez, 2010):

• The perimeter walls are built in brick or prefabricated blocks, topped with a reinforced concrete beam.

• The panels of the tendon wall should have a rectangular shape, with the columns being placed and embedded, and then the lower and upper beams or girders are placed, forming the structural framework.

• The columns of the confining frame structure of the tendon wall can be sized from 0.75 metres to a maximum of 1.2 metres, with an average height of 2.4 metres maximum.

• Having the structural framework ready, the laying of the barbed wire tendon begins, which should have a symmetrical vertical spacing between 20 and 30 centimetres, tensioning and intercalating its arrangement of the distance between the respective vertical pars.

• The fique sacking begins to be placed on the structural framework.

• Subsequently, the electrical and hydraulic installations of the house are carried out.

Figure No. 3.7

Geometry of the tendon wall system

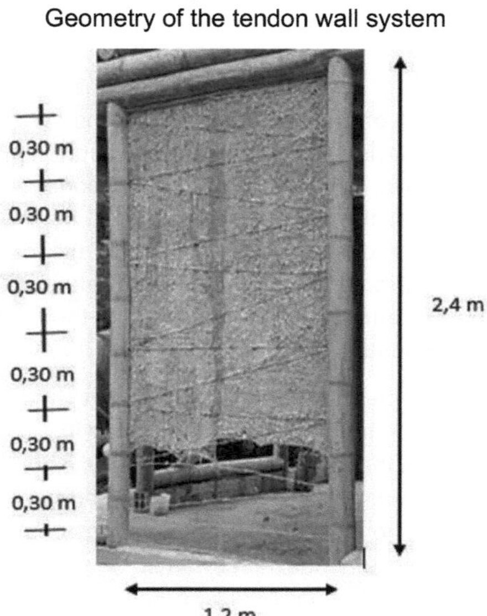

Note: Taken from https://dokumen.tips/documents/3-muros-tendinosos-1.html

After building the constructive skeleton of the tendon wall, following and complying with the above steps, the mortar mixture is prepared and placed on the tendon wall as a plaster, which consists of the application of several layers of mortar, with the plasticity and consistency of the mortar. necessary to the mortar mixture, in order to achieve adhesion and subsequent hardening, to act in a monolithic manner. The first layer of the plastering or chamfering process is applied with force on the surface of the constructive skeleton of the tendon wall, it should be left to set or dry for several hours, to apply the respective layers until the surface is completely smooth and uniform, thus concluding the construction of the tendon wall, (Figure No. 3.8).

Figure No. 3.8

Construction of the tendon wall system

Note. Taken from: http://www.zuarq.co/

3.3.3.- Structural behaviour of Tendon Walls. From the constructive point of view, the sinewy wall has the behaviour of a portico-style diaphragm with heterogeneous and anisotropic characteristics, where its material components present a non-linear mechanical behaviour, due to the different physical and mechanical characteristics, i.e., they vary according to the direction of application of the forces to which they are subjected. In conclusion, given the heterogeneous nature of the materials, their structural behaviour not the sum of the behaviour of all their components. Individual analysis of the main components of the tendon wall is performed, Figure No. 3.9:

- Structural frame columns, their main function to confine and support the other components of the wall, and to transmit the vertical loads to the foundation wall.
- Barbed wire, its function is that of a tendon or cable, its diagonal placement is with the purpose of being able to support and counteract the horizontal tension loads acting on the tendon wall, without presenting deformations or breakage. The sharp, fixed spikes or protrusions placed transversally have the function of supporting the sack of sacking and the mortar that has already set.
- Fique sack, its purpose is to adhere the mortar layer to its surface, it can support some perpendicular loads of seismic character to its plane.

Figure No. 3.9

Comportamiento Estructural del Sistema de muro Tendinoso

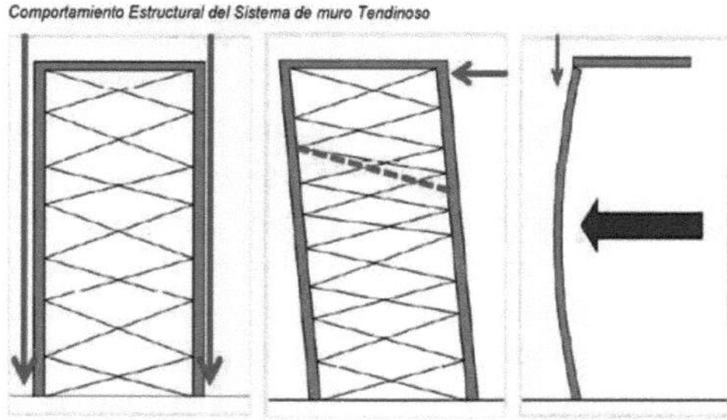

Note: Taken from https://dokumen.tips/documents/3-muros-tendinosos-1.html

- Mortar mixture, produces confinement to provide bonding between its elements and rigidity to the framework, constituting a monolithic structural element.

An important particularity of the tendon walls is their good behaviour and response to seismic forces. Due to their low unit weight, the seismic loads acting on the masses of the elements are lower, as is the acceleration produced on the structure, producing a lower seismic effect on lateral displacement. Velázquez confirms this assertion, commenting that in the earthquake that occurred in Armenia (Colombia) in 1998, the behaviour of the houses built with tendon walls was optimal, due to the good ductility of these constructions. The other masonry construction systems present in that locality did not have the same behaviour, as there were many collapses accompanied by loss of human lives, (2010, p. 9).Mora, in his laboratory research, affirms that the Thomas type tendon wall system is the most flexible of the tested walls, also valuing its maximum load-bearing capacity of 2.42 kN (246.8 kg), recommending this type of construction as dividing elements in buildings, (2022, p. 105).

CHAPTER IV

ANALYSIS OF THE SINEWY WALL CONSTRUCTION SYSTEM

It presents, for information purposes, the contextualisation of the different analyses and points of view made by a number of researchers interested in this new constructive theme.Thomas in his investigative character maintains that, "the investigative bet consisted of stimulating an integral approach to environmental design, which recreates the positive aspects of the existing technoculture" (2002, p. 28). 28), where a political demarcation prevails that comes from the conquest and later colonisation, as architecture with durable materials such as brick, which he calls colonial technoculture, was imposed on the local biodegradable technoculture of wood (malocas and bahareque) which was excluded being of indigenous type, which he calls the great oblivion technological education. From this ideological perspective, the tendinous system is materialised, developing construction techniques and using traditional wooden structures of the region, so that the inhabitants were their self-builders through the ancestral customs of the mingas and convites (indigenous tradition of community work), other important factors being that the resulting constructions are comfortable for their inhabitants (Figure 4.1). Thomas defines it as "appropriate technology" when designing an environmentally friendly construction system that responds to the socio-cultural determinants of the place, from an aesthetic, structural and economic point of view, (2002, p. 31).

❖ Guerrero y Casas, analyses the seismic aspect, where he emphasises that the Pacific region is located in a zone of high seismicity, the behaviour of the different structures and their vulnerability to these events and the safety that the constructions can guarantee in the face of earthquakes, adding that the

The 1999 Armenia earthquake in 1999 highlighted the high seismic vulnerability of the buildings of less than one and two storeys that were built without complying with the requirements of seismic-resistant design, (2002, p. 51). This system can be classified as confined load-bearing walls and its basic structural system consists of a timber, guadua or steel frame, which when integrating tendon wall panel to this structure works as a load-bearing wall, this construction system has been successfully used in one and two-storey dwellings, (Guerrero and Casas, 2002, p. 52-53).

Figure No. 4.1

Construction of the tendon wall system in steel structure

❖ Velázquez carries out an analysis of the tendon wall system:

Due to its versatility, speed construction and economy, it was a pilot project of the National Federation of Coffee Growers of Colombia (Federecafé) using this construction system in the Department of Valle del Cauca (Colombia) and specifically in the municipalities of the coffee axis (Caicedonia, Sevilla, Restrepo, Trujillo, La Victoria, Bolivar) among others, where there are large plantations of bamboo or bamboo, (Velázquez, 2010, p. 8-10), Figure No. 4.2.

Figure No. 4.2

Progress on the construction of a house with the sinew wall system

This construction system is not endorsed by the Colombian Seismic Resistant Construction Regulation, (NSR-10), Title G, however it had its first fire test during the Armenia earthquake in 1999 (seismic magnitude on the Richter Scale 6.1) where it demonstrated optimal behaviour due to its inherent ductility that prevents the collapse of the walls, some of the houses built by this system and that were affected by the earthquake did not present major damage nor did their walls collapse, unlike the bad behaviour of traditional brick construction systems, (Velázquez, 2010, p. 8-10). 8-10). Due to its ease of construction, speed of execution and low cost, this construction system has made it possible to respond to the needs of many people with low economic levels. These characteristics have made the tendon walls popular in these areas, reproducing and adapting the construction methods with new materials and various geometric configurations (Velázquez, 2010, p. 15-16). From the structural point of view, the tendon wall is heterogeneous, it is a composite material, anisotropic and its components present a non-linear mechanical behaviour, its structural characteristics are the following:

- Confining structure, these elements work with a double structural function by transmitting the vertical loads to the foundation, horizontally confining the panel frame, as well as being the physical support for all the other elements of the system.
- Although it does not have a specifically structural function, its main function is to give consistency to the wall in the face of loads perpendicular to its plane of placement, it also gives adherence to the tendon wall as a whole.
- Barbed wire, its behaviour is that of a tensioning cable that absorbs tensile loads in the horizontal direction of the wall, it also allows the grip and support of the natural fique lattice.
- Fique natural fibre lattice is the surface that allows the placement and cohesion of the mortar in the face of loads perpendicular to the plane of the wall, (Velázquez, 2010, p. 21-22).

Additionally, Velázquez differentiates the structural systems of the tendon wall with their various constituent materials, Figure No. 4.3:

- Structure with metallic profiles, the sections of these metallic profiles A-36 is small (1 1/2" to 2" x 1/8"), its anchorage to the foundation is totally embedded (the angle or profile is embedded) or its fixation can be made by the anchorage system with embedded bolts, the union between the elements of the framework is made by means of electric welding, This means that it can assume movements of the structure or framework without compromising the stability of the wall. Galvanised hooks must be welded to these profiles to fix the barbed wire tendons, as the sections of the profiles are small, it does not reduce the architectural area of the houses.

- Timber structure, the sections of this framework in sawn or round timber can be 4" x 4" x 6 metres, its anchoring to the foundation is totally embedded (the column is embedded), the union between the elements of the framework is made by means of bolts, it is required that wood has undergone a process of immunisation.

- Guadua structure, the sections of this framework can be from 10 centimetres to 15 centimetres in diameter, its anchoring to the foundation is totally embedded (the column is embedded), the union between the elements of the framework is made by means of bolts, it is required that guadua has undergone a process of immunisation.

Figure No. 4.3
Construction of one-storey housing complexes with the tendon wall system

The geometry of the sinew wall panels is rectangular in shape, the columns of the confining structure can be sized from 0.75 metres to a maximum of 1.2 metres, with an average height of 2.4 metres maximum, the placement of the barbed wires can be between 20 and 30 centimetres, interspersing their distance between the parallels, (Velázquez, 2010, p. 24- 26). The functionality of the tendon wall system is analysed from the following point of view constructive view, as a whole of the dwelling itself, highlighting the following functionalities, Figure No. 4.4:

Figure No. 4.4
Construction of dwellings with the sine-wall system

- Enclosure, its basic function is to separate the different rooms of the dwelling, it constitutes a solid matrix as a whole that is capable of resisting impacts perpendicular to its plane of action and is relatively light in comparison with conventional brick masonry construction systems.
- Resistance, the tendon wall is not self-supporting and its vertical structure must be embedded in the foundation, the resistance of the whole system of the tendon walls

is subject to the confinement given in the structuring of the construction system, being suitable for constructions of one and two floors.

- Insulation, it does not have any kind of acoustic or thermal insulation.

- Durability, the tendon wall requires proper maintenance when there are cracks, flaking mortar, cracks in the wall, cracks in the wall and cracks in the wall itself.

The durability of the construction joints can be ensured by timely repairing.

- Watertightness, the tendon wall by its nature not watertight due to the porosity of the mortar, to avoid this phenomenon the wall must be waterproofed on the outside to prevent leaks.
- Construction time, one of the most important advantages of the tendon wall system is its fast construction process.
- Versatility, this construction system is applicable to social housing (VIS), in country houses of high stratum, for its adaptability to all kinds of projects with different climates and topographies of the terrain.
- Aesthetically, constructions with tendon walls have a lot of freedom from an architectural point of view, (Velázquez, 2010, p. 42-44).

❖ Franco (2019), analysed different traditional construction techniques in Colombia, mainly those based on the use of bamboo in the

The project also studied the possible bioclimatic criteria to be taken into account in order to guarantee an optimal adaptation to the local climate, resulting in the comfort of the users without the need for high energy consumption. This system gradually became widespread in other regions, mainly due to its ease of construction, as it does not require skilled labour and can be carried out by community self-construction methods.

❖ Bedoya emphasises that the aesthetic aspect of the structure is very necessary for the acceptance or rejection of a construction technique or material.

of construction, (2011, p. 82). It lists the characteristics, properties and functions of the materials involved in the construction of a tendon wall, which are the following elements:

- Barbed wire, base material that acts as reinforcement in the tendon lattice, its structural function is to absorb the tensile stresses in the wall panel.

- This material is the core of the panel and composed of sacks or bundles of fique, it acts as a mesh that covers the entire length of the panel. Its main function is to give adherence to the tendon system with the mortar that is loaded in successive layers until the desired thickness of the wall is obtained.
- Mortar of glue, this material is the one in charge of giving solidity to the panel of the tendinous wall and at the same time giving it its final finish, the activities of the placement of this mortar of glue is similar to the plaster, the first activity is the loaded of the panel by means of a mixture of mortar rich in cement and water that is thrown on the framework of natural fibre with the purpose of obtaining adherence, later after

the drying of this layer successive layers are placed that will give the necessary thickness to the tendinous wall.

- Confining structure, which is composed of vertical elements (columns) and horizontal elements (beams) that will give rigidity to the wall system, the materials can be wood, guadua, angle iron, reinforced concrete, (Bedoya, 2011, p. 82-83).

❖ Builes analyses the constructive aspect of the tendon wall, where the foundation consists of an overlying beam, on top of which a perimeter masonry wall is built with a brick starting point with the

In order to provide insulation to the structural system and the tendon wall itself, the panels that make up the resistant structure are built on this wall and are made up of columns and beams that provide confinement to the system, the columns are embedded in a sill both at the top and bottom in order to provide rigidity to the system, (Builes, 2011, p. 44-45).

❖ Casas defines it as a non-conventional construction system, which must provide safety, stability and durability over time, (2011, p.

27). It defines the aspects of this system to be evaluated:

- Environmental aspect, this is composed of two orders:

1. Formal order is the urban or rural environment where the system will be implemented.
2. Functional order, which is composed of geography, climate and building regulations.

- Aspect technological, this includes the level of usage of the level of use of the building system.
- Socio-economic aspect, relates the economic situation of the region, (Casas, 2011, p. 28-36).

❖ In their research, Giraldo - Raigoza and Sanchez, emphasises the construction of: Housing in the rural area with the application of bioarchitecture of tendon wall, based on the characteristics of the social environment, areas of influence with latent need for a model of improvement at the level of housing and need, it is considered urgently to meet the deficit presented. Environmental awareness is another of the factors to be addressed in rural areas from the incursion of the new rural housing model proposed (2016, p. 16). Guadua angustifolia is the most important raw material in the construction of a kind of sinewy wall, it is a renewable and sustainable resource as it fixes carbon dioxide (CO_2), complying with constructive sustainability.

❖ Mora carries out experimental research in the laboratory, manufacturing various types of tendon walls subjected to lateral loads,

The dimensions of these tendon walls have a height of 2.20 metres, a width of 1.20 metres and a thickness varying between 5 and 7 centimetres, listing the following prototypes, (Mora, 2022, p. 47-49):

► Thomas type tendon wall, which consists of a guadua portal, an internal matrix in fique sacking, barbed wire and coated on both sides.mortar faces with cement and sand.

► Chacon type tendon wall, consisting of a guadua portico, an internal matrix of wire mesh with vein, internal grid reinforced with 3/8" corrugated steel and coated on both sides with mortar with cement and sand.

► Morachá type tendon wall, which consists of a guadua portico, a internal matrix of guadua matting, reinforced with 3/8" corrugated Steel and coated on both sides with cement and sand mortar.

From the bending test with lateral loads carried out on the tendon wall samples in the laboratory, following results for the maximum ultimate load are obtained, (Mora, 2022, p. 91):

► Thomas type tendon wall, X= 6.12 kN.
► Chacon type tendon wall, X= 7.00 kN.
► Morachá type tendon wall, X= 6.55 kN.

	CARGA MÁXIMA DE ROTURA		
	THOMAS	CHACÓN	MORACHÁ
Ensayo 1	6,75	8,2	6,9
Ensayo 2	5,99	6,15	6,64
Ensayo 3	5,4	7,55	6,47
Ensayo 4	6,33	6,08	6,2
X =	6,12	7,00	6,55
S =	0,5705	1,0506	0,2941
CV =	9,33	15,02	4,49

When performing a statistical analysis of the values of the lateral bending test of the tendon walls, the highest arithmetic mean corresponds to the Chacón test, the Thomas value is 12.6% lower and the Morachá value is 6.4% lower, with respect to the highest value. The value of the coefficient of variation (CV) of the test of Chacón has a high variation (15.02%), followed by the Thomas test (9.33%) and the lowest coefficient for the Morachá test. It is concluded that the various types of tension walls perform well under horizontal or lateral bending loads.

4.1.- Discussion and Conclusions on Tendon Walls. A series of discussions and conclusions are presented on the tendon walls, which represent a novel, non-conventional construction system.

• Tendon systems are considered to be constructive forms that have their origin in the zoological architecture of vertebrates, such as the combination of skeleton, muscles and tendons (Cardellach, 1970), in relation to this research, this constructive system proves to be useful in elements of load and division, mainly in the direct relationship of the materials that make it up and the purpose for which they will be designed. This is supported by Cardellach (1970), who affirms that the only origin of these structural and constructive forms is at a higher level of mechanical sensitivity and natural inspiration. Cardellach concludes that the engineer, with this innovative spirit, must discover new forms of resistant frames for actual construction.

• The tendon wall system has been used in Colombia since the 1990s. As it is environmentally friendly, it has provided an answer to the need for low-income housing, as it does not require skilled labour. The confinement provided by the tendon wall system is highly resistant to gravity loads and very stable to seismic events, as was demonstrated in the Armenia earthquake (Colombia) in 1998 (Zuluaga, 2012, p. 14-15). From the technological point of view, the adoption of non-conventional alternative construction systems brings with it the innovation or improvement of these systems, either through the use traditional regional materials, the implementation of new construction, organisation and production methods to reduce construction costs, or the use of recycled materials in some cases, All these aspects listed are possibilities that at a given moment can become real and effective options for the use of existing resources, and in this way help the less favoured classes to improve their living standards with the construction of decent housing, such as those built in the neighbourhood of La Independencia, Municipality of Restrepo, Valle del Cauca, Colombia, Figure 4.5.

Figure No. 4.5

Urbanisation of houses built with the tendon wall system

The advantages of this tendon wall system are as follows:

• Ancestral peasant building traditions can be applied and improved.

• Labour and material resources from the same region are used.

• Mortar-coated panels can be given a rustic finish, or they can be stuccoed and painted afterwards.

• The panel system allows the internal installation of all kinds of pipes for electrical, hydraulic and gas installations.

• Due to their thickness, the panels serve as thermal and acoustic insulation in the house.

- The tendinous system is very economical so that houses can be built at low cost, and houses can be built in both rural and urban areas of a locality.
- The behaviour of this tendon wall system in the face of the seismic actions presented in the region was very positive, the 1995 Paez earthquake with a value on the Richter scale of 6.5, and the Armenia earthquake in 1998 with a value on the Richter scale of 6.4, qualified by their value as strong. The houses built with the tendon wall system did not suffer structural damage.
- The tendon wall system does not overturn in earthquakes, as was the case with wattle and daub and simple brick walls, and this seismic response factor is very important when it comes to safeguarding the lives of the occupants.
- It is an friendly construction system.
- This construction system is part of the framework set out in the Regional Action Plan for sustainable development of Agenda 21 of the United Nations, with regard to the recovery of traditional systems and materials, by using non-conventional and alternative systems.

The few disadvantages of this tendon wall system are:

- Direct dampness can occur in the walls located in the external parts of the dwelling, as a result of rainwater, which can be prevented by periodic maintenance with waterproof plastic paint coatings.
- When building in wood or guadua, these must be well protected against attacks by xylophagous insects, in order to guarantee their durability over time and good structural behaviour.
- Another very important factor to take into account is the constructive errors that may occur in its execution, as a result of inexperience in self-construction.

The following aspects are left as a topic for research and technical discussion:

- Such is the importance achieved by the tendon system in solving a structural strength problem that an original, economical, strong and elastic structure is obtained.
- Thomas materialises the tendinous system, developing construction techniques and using traditional wooden structures of the region, implementing the self-construction system.
- This tendon wall construction system is a response to the housing needs of the underprivileged classes, and can be implemented in any geographical location.
- With this new construction system technology, the aim is to comply with environmental requirements, taking into account, from this perspective, the application of new clean technologies in the field of construction.
- It is a basic tool that should guarantee a dignified quality of life, responding to the unsatisfied primary needs of low-income families.

• The community self-construction system can be implemented, as it does not require prefabricated elements.

• As it is a very flexible construction system, the architectural spaces of the house can be easily adapted and modulated.

• The system is adaptable to any geographical area and topography, as well as being environmentally friendly.

• Because of its versatility construction, it is possible to use various materials in the structure of the house, be it wood or guadua from the region.

The following conclusions are drawn from this research process:

1. The feasibility of the construction system to address the housing shortage in both urban and rural areas was demonstrated.
2. The tendon system results in a quality construction with local materials.
3. From its philosophical principle it contributed the integrative concept of appropriate technologies.
4. From a professional and academic point of view, working with the community provides an insight into ancestral constructive knowledge.
5. Academically, the concept of exclusionary architecture is overcome by the evolutionary creative function of inclusive architecture.
6. Philosophically, the social acceptance of this technology is significant, with the community rating it as appropriate for them to implement.
7. The construction projects of the tendon walls in rural or urban developments must satisfy the physical and basic social needs of the communities, in accordance with sustainable construction and maintaining a balance between the existing ecosystems.

BIBLIOGRAPHICAL REFERENCES

Arredondo, F. (1963). Estudio de Materiales, III.-Cales. Madrid: Instituto Eduardo Torroja de la construcción y del cemento.

Bedoya Montoya, C. M. (2010). Tendinous walls in architecture and construction. Book four of architecture (1ª ed.). Medellín: Facultad de Arquitectura, Universidad Nacional de Colombia, sede Medellín.

Bedoya Montoya, C. M. (2011). Construcción Sostenible, para volver al camino. Biblioteca Jurídica DIKË, UNESCO Chair in Sustainability. https://issuu.com/iscucen/docs/construcci n_sostenible_para_volver

Builes Hoyos, T. and Giraldo Montoya, C. (2011). State of the art of guadua as an alternative material for sustainable construction. [Degree thesis. Universidad EAFIT]. https://repository.eafit.edu.co/handle/10784/5451

Cardellach, F. (1910). Filosofía de las estructuras. Barcelona: Librería de A. Bosch. https://bibliotecadigital.jcyl.es/es/catalogo_imagenes/grupo.do?path=1046581 7

Casas F. L. H. (2011). Sostenibilidad, Sistemas Constructivos, Muros Tendinosos, Seminario Biocasa - CAMACOL, Hábitat y Desarrollo Sostenible, Santiago de Cali, Colombia.

Comas, J. (1977). Introduction to general prehistory (3ª ed.). Mexico: Colmex Ltda.

Cruz A. J. C., Cardona G. J. C., Hernández P. D. M. (2013). Innovation in Engineering for the Construction of Eco-Sustainable Houses. Seminar on Innovation in Research y Education at Engineering. Cartagena, Colombia. https://antiguo.acofipapers.org/index.php/acofipapers/2013/paper/viewFile/15 7/59

De Roux, R., R., (1990). History of humanity (2ª ed.). Bogotá: Estudio.

Fernández, P. D. (2016). Composition and structure: Revaluation of the constructive technique of stacking as a design strategy in contemporary architecture. Doctoral thesis Faculty of Architecture, Planning and Architecture.

Design, Universidad Nacional del Rosario, Argentina.Retrieved 16 February from 2018 From: http://www.fapyd.unr.edu.ar/wp-content/uploads/2017/09/09/tesis_fernandez_paoli.pdf.

Gaiker IK4 Research Alliance (2007). Materials Recycling: perspectives, technologies and opportunities. Department of Innovation and Economic Promotion, Bizkaiko Foru Aldundia, Spain.

Gama, J., Cruz, T., Pi-Puig, T., Alcalá, R., Cabadas, H., Jasso, C., et all, (2012). Earthen architecture: adobe as a building material in pre-Hispanic times. Bulletin of the Mexican Geological Society, 64 (2), 177-188.

García, D. A. F. (2019). Analysis of a construction with bamboo and adobe. Proposal for the construction of a school in Quibdó (Chocó, Colombia). Office of Environment (OMA-UDC) Vicerreitoría de Economía, Infraestruturas e Sustentabilidade,53.

https://ruc.udc.es/dspace/bitstream/handle/2183/25606/Premio_UDC_sustentabilidade_TFG_TFM_I_2018.pdf?sequence=8#page=53

García León, D. F. and Velásquez Ramírez J. A. (2021). Feasibility of tendon walls based on PET strips and banana rachis fibres. [Degree thesis, Universidad de La Salle]. https://ciencia.lasalle.edu.co/ing_civil/959/

Giraldo M. S., Raigoza, V. A., C., and Sánchez Restrepo, A., (2016). Feasibility study for the construction and marketing of ecological housing of priority interest using the tendon wall technique in a rural area of Risaralda. [Degree work, Universidad Tecnológica de Pereira]. https://core.ac.uk/download/pdf/84108257.pdf

Ghoreishi, K. K. (2011). Ecomaterials and sustainable construction. Canada: Creative Commons.

Guerrero, P. and Casas, F. L. (2002). Materiales y sistemas alternativos para la vivienda, Los muros tendinosos. Revista CITCE, Territorio, construcción y espacio. 4 (Jul/Dec), 48-55.

Hernández, N. J. (2013). The language of structure: the decomposed wall. [Final Work of Master's thesis, University Ramon Llull]. https://www.recercat.cat/bitstream/handle/2072/256777/Hernandez-Navarro-MPIA.pdf?sequence=1

Hidalgo, L. G. (1978). Nuevas técnicas de construcción con bambú. Bogotá, National University of Colombia. Publisher, Estudios Técnicos Colombianos.

Howell, F. C. 1982. The human race. Revista de Occidente, Extraordinary IV (18-19), 21-42.

Instituto Sindical de Trabajo, Ambiente y Salud (ISTAS). (2005). Guide to Sustainable Construction. Ministry of the Environment, Spain. https://istas.net/descargas/CCConsSost.pdf

Keyser, C. (1982). Materials science for engineering (1ª ed.). Mexico: Limusa S. A.

Lamuz, F., and Andrade, S. (2015). Concreto reforzado, fundamentos (1ª ed.). Bogotá: Ecoe Ediciones Ltda.

Magaña, H. P. P. (2022). Production of recycled ecological materials from construction debris. CITAS Journal, Supplement (1). Vol. 8 (En/Jun), 70-92. https://revistas.usantotomas.edu.co/index.php/citas/issue/view/691

Management Solutions (2020). Managing the risks associated with climate change. https://www.managementsolutions.com/sites/default/files/publicaciones/esp/g estion-risgos-cambio-climatico.pdf

Ministry of Environment and Sustainable Development (MADS), (2022). Guía de materiales para la construcción sostenible, Dirección de Asuntos Ambientales, Sectorial y Urbana (DAASU). Bogotá D. C., Colombia.

Mora, Ch. W. F. (2022). Determination of lateral displacement in tendon walls under lateral, monotonic and cyclic loads. [Final work of Master's degree, University National University of Colombia]. https://repositorio.unal.edu.co/handle/unal/83329

Perdrizet, M. P. (1987). The men of prehistory. Bogotá: Norma S.A.

Pérez,M.A. (2014). Advanced applications of composite materials in civil engineering and building. Barcelona: 1ª edition Omnia Sciense. https://www.omniascience.com/books/index.php/monographs/catalog/view/77/301/466-1

Ramírez, Q. D. C. (2022). Rehabilitation techniques for masonry walls. [Degree thesis, Universidad Nacional Autónoma de México]. https://www.resilienciasismica.unam.mx/docs/TesisDianaRamirez.pdf

Rivas, R. I. (2009). Tendon walls. https://es.scribd.com/doc/214883597/3-Muros-Tendon walls-1.

Salazar A. J. (2012). Ecomaterials and VIS. A sustainable vision. https://www.colmayor.edu.co/wp-content/uploads/2019/10/5-materialesdeconstruccindebajo.pdf

Soufflot, J. G. (2012). Biography. http://www.wikiwand.com/es/Jacques-Germain_Soufflot0

Thomas, M. A. and Supelano, P. (2002). Diseño y Tecnocultura, Alternativas Constructivas: el Caso del Sistema Tendinoso, Revista Ingeniería y Competitividad, Facultad de Ingenierías, Universidad del Valle, Colombia, 4 (1), 25-32.

Thomas M. A., Supelano, P. and Vergara, C. (2015). Tendon + Tended = Tendinous, Revista digital Constructivo, 106, (April - May), 170-178. http://constructivo.com/cn/suscriptor/pdfart/150410043301_ARTICULO104.pd f

Torres, R. J. E. (2013). Bioarchitecture: Tendinous wall. [Blog Ingeniería en arquitectura y design environmental].

http://ingenieroenarquitecturamedioambiental.blogspot.com/2013/01/bioarqui etctura-tendinous-wall.html

Torroja M. E. (2010). Razón y ser de los tipos estructurales (7ª ed.). Consejo Superior de Investigaciones Scientific Research Council. Ediciones Doce Streets,S. L.file:///D:/INFO%20USER%20C%20NO%20BORRAR/Downloads/Razon_and_being_of_Structural_Types_E.pdf

Valenzuela, A. (2015). Las Patentes de Hormigón Armado. Dialnet, RITA, No. 3, April 2015. https://dialnet.unirioja.es/descarga/articulo/5094560.pdf

Velásquez R. J. A., and García L. D. F. (2021). Feasibility of tendon walls based PET strips and banana rachis fibres. Retrieved from https://ciencia.lasalle.edu.co/ing_civil/959

Velázquez R. A. (2010). Technology transfer: the case of tendon walls. [Master's thesis, Universitat Politècnica de Catalunya].

https://es.scribd.com/document/249974865/Velasquez-pdf

Vidaud, E. (2013). Construction and technology in concrete, From the history of cement.2013 (November), 20-24.
Vitruvii P. M. (1997). De architectura, opus in libris decem (1ª ed.), translated from the Latin by Oliver D., J. Madrid: Alianza Forma S. A.
https://web.seducoahuila.gob.mx/biblioweb/upload/Vitruvio_Polion_Marco.pdf

White, R., Gergely, P. and Sexsmith, R. (1980). Introduction to the concepts of analysis and design, Structural Engineering (Vol. 1). Limusa S. A.

Zuluaga, C. and Zuleta, A. (2016). Arquitectura sostenible, Sistema tendinoso, Revista digital. Colconstrucción, February (14-15).
https://issuu.com/colconstruccion/docs/colconstruccioned3-5_imprimir

MIX
Papier aus verantwortungsvollen Quellen
Paper from responsible sources
FSC® C105338

Printed by Books on Demand GmbH, Norderstedt / Germany